ART OF SOUND

C000002749

Bringing together a diverse group of world leading professionals across Post-Production Film Sound and Electroacoustic Music, *Art of Sound* explores the creative principles that underpin how sonic practitioners act to compose, tell stories, make us feel, and communicate via sound. Revealing new understandings through analysis of interdisciplinary exchanges and interviews, this book investigates questions of aesthetics, perception, and interpretation, unveiling opportunities for a greater appreciation of the artistry in sound practice which underpins both experimental electronic music and the world's leading film and television productions.

It argues that we can better understand and appreciate the creative act if we regard it as a constantly unfolding process of inspiration, material action, and reflection. In contrast to traditional notions, which imagine outputs as developed to reflect a preconceived creative vision, our approach recognises that the output is always emerging as the practitioner flows with their materials in search of their solution, constantly negotiating the rich networks of potential. This enables us to better celebrate the reality of the creative process, de-centring technologies and universal rules, and potentially opening up the ways in which we think about sonic practices to embrace more diverse ideas and approaches.

Art of Sound provides insight into the latest developments and approaches to sound and image practice for composers, filmmakers, directors, scholars, producers, sound designers, sound editors, sound mixers, and students who are interested in understanding the creative potential of sound.

Andrew Knight-Hill is a composer and researcher creating works that explore the interface between music and sound practice. His works have been heard internationally in film festivals and contemporary music concerts. He recently composed music and sound design for the immersive theatrical event "Over Lunan", which was nominated for Outstanding Cultural Event as part of the 2022 Thistle Awards for Scottish Tourism.

He is Associate Professor in Sound Design and Music Technology at the University of Greenwich; leader of the SOUND/IMAGE Research Centre; co-director of the Loudspeaker Orchestra Concert Series; and convenor of the annual SOUND/IMAGE conference.

www.ahillav.co.uk

Emma Margetson is an acousmatic composer and sound artist. Her research interests include sound diffusion and spatialisation practices; site-specific works, sound walks and installations; audience development and engagement; and community music practice. She has received a variety of awards and special mentions for her work, including first prize in the prestigious L'Espace du Son International Spatialisation Competition by INFLUX (Musiques & Recherches), klingt gut! Young Artist Award in 2018 and Ars Electronica Forum Wallis 2019.

She is Senior Lecturer in Music and Sound at the University of Greenwich, a Research Fellow for the AHRC project "Audiovisual Space: Recontextualising Sound-Image Media" and co-director of the Loudspeaker Orchestra Concert Series.

www.emmamargetson.co.uk

Sound Design
Series Editor: Michael Filimowicz

The *Sound Design* series takes a comprehensive and multidisciplinary view of the field of sound design across linear, interactive and embedded media and design contexts. Today's Sound Designers might work in film and video, installation and performance, auditory displays and interface design, electroacoustic composition and software applications, and beyond. These forms and practices continuously cross-pollinate and produce an ever-changing array of technologies and techniques for audiences and users, which the series aims to represent and foster.

For more information about this series, please visit: www.routledge.com/Sound-Design/book-series/SDS

Art of Sound
Creativity in Film Sound and Electroacoustic Music

Andrew Knight-Hill and Emma Margetson

Routledge
Taylor & Francis Group

LONDON AND NEW YORK

Designed cover image: © Emma Margetson & Andrew Knight-Hill

First published 2024
by Routledge
4 Park Square, Milton Park, Abingdon, Oxon OX14 4RN

and by Routledge
605 Third Avenue, New York, NY 10158

Routledge is an imprint of the Taylor & Francis Group, an informa business

British Library Cataloguing-in-Publication Data
A catalogue record for this book is available from the British Library

Library of Congress Cataloging-in-Publication Data
Names: Knight-Hill, Andrew, author. | Margetson, Emma, author.
Title: Art of sound: creativity in film sound and electroacoustic music /
Andrew Knight-Hill and Emma Margetson.
Description: Abingdon, Oxon; New York: Routledge, 2023. |
Series: Sound design |
Includes bibliographical references and index.
Identifiers: LCCN 2023019419 (print) | LCCN 2023019420 (ebook) |
ISBN 9780367755881 (paperback) | ISBN 9780367755898 (hardback) |
ISBN 9781003163077 (ebook)
Subjects: LCSH: Film soundtracks—Production and direction. | Motion picture
music—Production and direction. | Electronic music—Production and direction. |
Soundscapes (Music)—Production and direction. |
Creation (Literary, artistic, etc.) | Film composers—Interviews.
Classification: LCC ML2075 .K57 2023 (print) |
LCC ML2075 (ebook) | DDC 781.5/42092—dc23/eng/20230614
LC record available at https://lccn.loc.gov/2023019419
LC ebook record available at https://lccn.loc.gov/2023019420

ISBN: 978-0-367-75589-8 (hbk)
ISBN: 978-0-367-75588-1 (pbk)
ISBN: 978-1-003-16307-7 (ebk)

DOI: 10.4324/9781003163077

Typeset in Optima
by codeMantra

Access the Support Material: www.routledge.com/9780367755881

To all who listen.

Contents

x   Contents

Diagrams

Acknowledgements

Any undertaking is always a collaborative activity. We are incredibly grateful for a wide range of support in many forms, which made this book possible.

Support from the Arts and Humanities Research Council (AHRC) gave us the space and time to undertake this project, via an Early Career Researcher Leadership Fellowship grant ("Audiovisual Space: Recontextualising Sound-Image Media" – Grant reference: AH/V006975/1). In addition to the financial support, we are incredibly grateful to the generosity of the AHRC teams and their Leadership Development programme which equipped us with the confidence to realise the fullest ambition of this project and to Prof. Miguel Mera who acted as a mentor providing guidance on both the delivery of this research and the wider questions of research leadership.

Each of the contributors to this volume gave their valuable time generously to the project, often engaging in deep and expansive conversations that far exceeded our initial intentions. Our analysis and conclusions build upon their incredible insights, wisdom, and knowing through doing, and we are so incredibly thankful to each of them: Ann Kroeber, Paula Fairfield, Randy Thom, Nina Hartstone, George Vlad, Steve Fanagan, Jesse Dodd, Onnalee Blank, Peter Albrechtsen, Hildegard Westerkamp, Annette Vande Gorne, John Young, Trevor Wishart, KMRU, Annie Mahtani, Nikos Stavropoulos, and Natasha Barrett.

The opportunity to engage with so many world leading professionals would never have been possible without the support and friendship of Paula Fairfield. A 2018 collaboration between Paula, composer Brona Martin and Andrew Knight-Hill (*Immersive Hyperreal Soundscapes*) served as an early prototype for the dialogues and exchanges in this volume. Paula's contribution to the project has been invaluable, from consulting on the initial application for AHRC support, through to the practical realisation, connecting us to so many of the incredible Film Sound professionals featured in this volume. We are in little doubt that her own enthusiasm and generosity, not to mention her own status as an outstanding professional, were essential in convincing many of the contributors to participate in this research project.

Special thanks are also due to Andrew Wilson, Chair of the Association of Motion Picture Sound (AMPS), for making introductions and connecting us to Nina Hartstone and Peter Albrechtsen, and Karol Urban of the Cinema Audio Society (CAS) for introducing us to Jesse Dodd.

Outside of the direct scope of this book, we are grateful to our colleagues in the SOUND/IMAGE Research Centre for their continued collegial support and mutual inspiration, especially that of Prof. Stephen Kennedy for his insightful mentoring, detailed reflections and critique upon earlier drafts of this manuscript.

Andrew Knight-Hill & Emma Margetson,
June 2023

1 What, Why and Who

Introduction and Context

This is a book about how sounds communicate, how they make us feel, and how they tell stories. Revealed through conversations with world leading creative professionals working in the UK, US, and Europe, we unveil hidden spheres of creativity by exploring the creative artistic processes that underpin some of the world's leading Film Sound and Electroacoustic Music.

Our goal is to better understand 'how' and 'why' creative decisions are made, to provide new ways of thinking and talking about how sound practices operate, and celebrate the artistry that exists in the field. Through interviews and discussions with sound practitioners, we have worked to highlight making as a process (a fluid act of doing) which offers up opportunities for new interpretations of practice that might encourage innovation and exploration, de-centre technological and descriptive approaches to understanding[1] and demonstrate the fundamentally artistic nature of sonic practices. We hope that this book can provide new pathways of access for those seeking to work creatively with sound, fresh opportunities for existing practitioners to reflect upon and think about their art, and provide a platform for potential collaborators to more meaningfully understand the rich potential of sound.

The parallel worlds of Film Sound and Electroacoustic Music share a common medium (sound), and many of the same tools and processes (digital audio workstations, plug-ins, microphones, etc.). But there has been a chasm of communication between these sister fields. Our research has sought to bridge this divide, bringing these two worlds together in dialogue; seeking to compare approaches, insights, and inspirations, to reveal where the two worlds might speak to one another in mutually beneficial ways.[2]

We hope that this book encourages more exchange, collaboration, and dialogue, such that it inspires all those who are already passionate about sound to discover something new in the familiar, and that it intrigues those yet to discover the power of sound, to awake and hear its full potential.

1.1 What Will You Find in This Book?

This book contains a mixture of individual interviews with practitioners, dialogues between pairs of professionals in Film Sound and Electroacoustic Music, and chapters discussing the trends and themes that emerged from within these many conversations. What is important to note at this start is that it is not a study *on* these practitioners, it is a study *with* them.

DOI: 10.4324/9781003163077-1

We've structured the book into a number of distinct sections:

- **What, Why and Who: Introduction and Context** – An explanation of why we undertook this research, our goals, and the wider context of the research (including definitions of what we mean when we talk of Film Sound and Electroacoustic Music).
- **Making Sense of Making** – An explanation of the methods we used and the questions we asked.
- **Mapping Stories of Creative Sonic Practice** – An in-depth exploration of the main themes emerging from our interviews and dialogues, culminating in Our Map of Sonic Creativity.
- **Interviews** – One-to-one interviews with world leading professionals in Film Sound and Electroacoustic Music, exploring their individual creative practices and inspirations.
- **Dialogues** – Guided conversations between Electroacoustic Composers and professionals in Film Sound, colliding knowledges and approaches to reveal synergies and divergences.
- **Celebrating the Art of Sound** – A summary of key research findings with suggestions for their tangible application and future potential.

We invite readers to chart their own course through this book. One might jump straight into the interviews and discussions with practitioners, seeking to draw out personal insights; Or choose to explore *Our Map of Sonic Creativity*, exploring the trends that we have uncovered via our own reflections and analysis. We hope that you will discover practitioners who are new to you, as well as re-discovering, or knowing in new ways, the workflow of practitioners with whom you may already be familiar. In all cases, we hope that this book stimulates thought, reflection, and response to questions of sonic creativity.

Before we dive into the details of our study, it is important to clarify:

1.2 What Do We Mean by Film Sound?

Film Sound, as we know it today, emerged out of the confluence of 19th century artistic practices of theatre sound, vaudeville, and opera, brought together with the new art of moving pictures. Technological innovations in sound recording enabled ever greater potentials for the synchronisation of sound with moving image, settling with the widespread roll out of the optical soundtrack in the late 1920s. By 1929, almost all Hollywood movies featured synchronised soundtracks, and sound practitioners were engaged as key artists within the collaborative process of film making (see Hanson 2017). Subsequent decades have seen significant developments in sound recording, manipulation, and reproduction technologies (including stereophonic sound and surround sound), but a relative erosion of the appreciation for sound's powerful potential to tell stories.

As Michel Chion notes, one of the challenges of Film Sound is that it is frequently defined in relation to this historical moment of transition from silent film to talkies, which instantly positions it as an addition, "a film without sound remains a film; a film with no image, or at least without a visual frame for projection, is not a film" (Chion 1994, p.143). The natural extension of this idea is to devalue sound's importance. How can an addition be regarded as truly significant to the storytelling potential of film? If a thing is not essential, how can it be vital?

It is true that Film Sound is created alongside a visual component, and that this often affords a structural frame with which the sound operates, but this is not to say that the visual is limiting to sound on the contrary, one of the key goals of this study is to reveal, and elaborate the detailed artistry of sonic practices so as to demonstrate the very real potential that sound has to benefit the work of Filmmakers, Game Designers and Artists.

When we talk about Film Sound in this book, we're actually referring to Post-Production sound.[3] Post-Production sound encompasses the diverse practices of sound editing, recording, and mixing that take place often after the principal photography of filmmaking has taken place. It is perhaps not widely appreciated beyond the Film Sound community that the majority of sounds heard within television programmes, films, and games are entirely constructed within the Post-Production phase. The actors' dialogue is often the only sound material from the production phase that makes its way through to the final film.[4]

Post-Production sound can be broadly divided into three main categories of practice – Recording, Editing, and Mixing – with specific specialist roles and subdivisions falling within each. As with any system of categorisation, there is fluidity between these categories and individual practitioners may undertake work across a number of areas. Post-Production sound roles include:

- Recording:
  - Sound Effects Field Recording – Capturing sound materials outside of the studio, such as soundscapes or specialist sound effects from objects or places.
  - Automatic Dialogue Replacement (ADR) – Re-recording lines of dialogue from principal actors to reflect late script changes or unusable recordings from the set and blending them with the original takes to mask the process of replacement.
  - Foley Recording/Mixing – Capturing performed movements and sound effects acted out by Foley Artists within a recording studio (often confused with Sound Effects Field Recording).

- Editing:
  - Dialogue Editing – Cleaning, aligning, and editing the multiple layers of voice recording captured during the production to deliver a clean set of takes. Includes layering in loop groups and additional voice elements for the background actors as well as editing the non-linguistic human utterances such as breath.
  - Sound Effects Editing – Editing and aligning the multiple layers of sound effects and atmospheres which create the depth and perspectives of the sound environments in film.
  - Sound Design – The creation of new sonic materials, either through synthesis or detailed editing of sound files, for spot effects or specialist objects or characters.

- Mixing:
  - Dialogue and Effects Mixers – Balancing the elements of various strands of material coming into the piece.
  - Re-recording/dubbing mixers – Blending all of the components of the soundtrack into one coherent and cohesive mix, these professionals work with the Director to manage the various component elements and deliver a final result.

Film Sound practitioners often work as part of a team, with special responsibilities for one (or more) of the aforementioned roles. Large teams are overseen by a Sound Supervisor, who will often work closely with the Director to realise their vision for the output. This involves directing the overall aesthetic approach of the various teams and professionals working under them, to sculpt a cohesive overall sonic result.

This collaborative nature of practice means that Film Sound professionals have a sense of identity formed as part of a larger collective. This was made explicitly clear in a number of interviews with participants and their deliberate use of plural pronouns, 'we', when discussing practical examples.

1.3 What Do We Mean by Electroacoustic Music?

Electroacoustic Music is an art of sound.[5] It treats all sound as musically valid material, radically extending the democratisation in sound that began with Schoenberg's 12-tone technique.[6] 'Electroacoustic' refers to a wide range of musical practices, each of which emanates from specific aesthetic and technical histories including contemporary sound art and electronic music composition. One early advocate of expanding musical possibilities was the composer Edgard Varèse who adopted the term 'organised sound' to describe his work. He became frustrated by repeatedly being drawn into debates with conservative traditionalists who refused to engage in open and constructive discussion of what music might be allowed to be.

> [To] stubbornly conditioned ears, anything new in music has always been called noise. But after all, what is music but organised noises? And a composer like all artists, is an organiser of disparate elements. [...] Our new medium [of sound recording] has brought to composers almost endless possibilities of expression, and opened up for them the whole mysterious world of sound.
>
> (Varèse in Cox and Warner 2002, p.20)

Varèse envisioned a new musical ideal in which all sounds were accepted as musically valid. He described music in terms of rhythms, frequencies, intensities, forms and layers, prioritising sonic texture (timbre) as a key musical parameter.

Electroacoustic Music developed significantly from the age of the phonograph, through magnetic tape to the digital age. Its practices have inspired and provided the impetus for many creative sound design and audio processing tools available today,[7] but while these tools are a physical manifestation of Electroacoustic Music's influence, the genre is often little known within sound design and music production circles. The practitioners featured in this book each have their own unique compositional voice, but fall within the following genre sub-categories:

- Acousmatic

  - Purely sound-based compositions, performed over a Loudspeaker Orchestra,[8] often with a focus upon the spectral and textural qualities of sound, so as to encourage heightened listening[9] and focused attention upon the qualities of sounds themselves.

- Soundscape

  - Compositions of sounds that create the sensation of experiencing a particular acoustic environment, perhaps created using the found sounds of an acoustic environment, either exclusively or in conjunction with musical performances.

- Ambient
  - A style of gentle electronic music that emphasises tone or mood to create fluid atmospheres of sound.

- Sound Arts
  - A diverse form of practices that lean more towards the gallery than the concert hall, which might include installations, sound sculptures, sound happenings, or fluxus performances.

Electroacoustic Music is often a highly individualistic practice, with one practitioner collecting all sound materials, making edits, mixing, and delivering the final result for their own performance over loudspeakers (perhaps via Loudspeaker Orchestra).

1.4 Why Are We Doing This?

This book seeks to realise a number of interrelated goals, which have the potential to change the ways in which we talk about, and therefore think about, sound practices.

Our aims and objectives are to:

- Reveal new insights through attention to the creative, communicative, and aesthetic processes of making.
- Encourage a greater appreciation for the artistry of Film Sound practices.
- Foster dialogues between Film Sound and Electroacoustic Music which help to challenge the traditional ways in which we think about these respective creative practices.
- Use these new insights and dialogues to further widen the diversity of creative professionals engaged in Electroacoustic Music and Film Sound.

1.4.1 A Focus on Making – Creative, Communicative, Aesthetic

Processes of creative learning and practice are often so bound up with the materials and actions of practices themselves that it can be difficult to extrapolate and communicate them outside of their sites of production. Knowing is achieved through doing, and much of the learning of Film Sound and Electroacoustic Music practice takes place within the studio via observation, listening, recording, and editing. It is often challenging to communicate underlying knowledge about practice because its explication is so contingent on the processes of making itself; it only emerges in the very act of doing. If we attempt to describe – steps taken, tools, processes, and software used – we gain an understanding of what was done, but know little about *why* those processes were chosen and how details were articulated. Such declarative forms of knowledge (knowing 'what') fail to get to the heart of the creative underlying ethos that drives the richness of expression through sound.

Applying approaches from anthropology (the study of human beings and their cultures), inspired by the works of Tim Ingold,[10] Georgina Born,[11] and Donna Haraway,[12] we have placed *making* at the heart of this research, seeking to understand the frameworks of tacit and embodied knowledges that underpin sonic practices, but which are often poorly reflected in existing studies and texts on the topic.

There are traditionally two main types of texts on the topic of sound practice:

1 Those written by creatives talking about *how* they made a particular work, which provides direct descriptions of processes and action, sharing anecdotes of specific projects

or sequences, discussing tools, technologies, and industry structures or workflows (e.g. Avarese 2017; Sonnenschein 2004). Scattered solo voices which inherently reflect only individualistic perspectives and risk making totalising conclusions about practice based only on partial individual truths.

2   Discussions of someone else's work (these may be written by fellow creatives or non-practitioners) which focus upon the final output of a project (e.g. a scene within a completed film) and which infer conclusions from analysis and discussion of the completed object (e.g. Donnelley 2005; Rogers 2014).

Both types provide what Tim Ingold calls – *knowledge about* – sound practices (Ingold 2013). They discuss creative works from an externalised perspective – ideas, theories, and tools – that might serve as a way of understanding the artworks but avoid discussing the underlying inspirations and creative ideas that drive the process of making itself. While such texts may contain some useful information, e.g., being introduced to practical pitfalls of how to manage your time, introducing types of cables and connectors etc. (e.g. Avarese 2017), they often apply rules and regulations (i.e. 'this is the way you do x') to make sense of practice. The danger of this approach is that everything can become formulaic and mechanical, leading towards lists of set rules which lead to the narrowing of possibilities and potential. Such approaches act to highlight the 'craft' of practice, while simultaneously denying the 'art'.

Our approach seeks to learn *with* practitioners, delving beyond descriptions of practical procedures to access *knowledge through* understanding their approaches to practice. We are seeking to learn more about the fundamental drivers and sensitivities which form creative and artistic sensibilities within sonic practices. We hope to celebrate *doing* as a valid form of knowledge, the investigation of which has the power to reveal new insights and instigate new dialogues around the practices of those working in sound.

1.4.2 Highlighting Artistry in Sound Practice

Music is generally considered as an artform in Western society, long held as a significant element in religious ritual and academic study.[13] It is sometimes even referred to as the ideal form of art, "the effect of music is so much more powerful and penetrating than that of any of the other arts, for [the other arts] only speak of shadows, but [music] speaks of the thing itself" (Schopenhauer 1819, p.333). This power stems from music being considered as abstract.

Sound is often considered the polar opposite of music, not even a shadow but something fundamentally concrete, earthly, and 'real'. It is a side effect of a source cause (e.g. car horn beep = car), which can convey some characteristics about that source (e.g. car horn beep = angry driver) but it always stands for something else, as part of our everyday experience of the world. This is the opposite of the aforementioned description of music; sound is 'not even a shadow' but something fundamentally concrete, earthly, and 'real'. In this framing, sound is a representation of a cause without the abstract potential of expression that music possesses. This denies sound permission to be considered as anything other than a side effect of action.

But, the technology to record and reproduce sound has enabled new possibilities of sonic creativity and fostered a growing attention and appreciation upon sound. Composers and musicians in the late 19th century began to utilise sounds from the everyday world in their compositions, and the rapid expansion of sound recording technologies after

1946[14] has only accelerated this function, leading to whole genres of electronic and contemporary music which utilise recorded sound as their primary musical material. Over its 80-year history, the field of Electroacoustic Music has sometimes struggled to be accepted as a form of music because it uses sound as its material.[15] Just as for Varèse, traditionalists can still reject the 'musical' potential of sound, but Electroacoustic Music has developed a rich vocabulary and discourse of thought that underpins and rationalises its practice. These narratives explore a wide range of facets of sound practice from the acoustic and technological, perceptual and psychoacoustic, to formal ideas about musical composition and structures. This enables Electroacoustic Music to challenge the limits of our definitions of what music can and might be. As a result of concerted efforts, sound-based musical practices are now firmly established within the contemporary music canon and often featured in high-profile concert series.

In the case of Film Sound, though there is an increasing appreciation of sound as a creative tool within film, it can remain devalued and considered less creative than other aspects of filmmaking. An indicator of the prevailing culture of value judgement is embodied within film credit rolls, in which the creative roles of image making (e.g. Cinematography, Costume Design, and Production Design) and narrative (e.g. Director, Script Writer, and Editor) are rewarded with top billing alongside key talent, while sound is often relegated to the tail end of the list, ranked as a technical role more closely aligned with the logistical elements of production.[16]

Technical film crafts such as Cinematography can draw artistic lineage from painting and photography, costume can draw artistic associations from fashion, production design can draw associations from sculpture, yet the traditional division between what is considered sound and what is considered music, denies sound its artistic association. Of course, the devaluing of sound is not only inherent to the film industry. Film Sound is devalued, in part because sound is devalued by Western society in general. And this denial of its creative potential means that Film Sound can sometimes be considered a craft practice, vocational rather than artistic. This framing, as a craft practice, also shapes the way in which the field is often conceptualised. It is a vicious cycle; because we have been taught to think about Film Sound primarily as a craft, the majority of texts and educational courses which deliver classes on Film Sound treat it as such, describing it in terms of a 'rules based' vocational frame. This, in turn, acts to re-enforce the notion of sound practice as a craft, while denying a complementary focus on the creative and artistic.

Electroacoustic Music has been able to overcome the devaluing of the sonic because music is valued by Western society as an art. Part of our mission, therefore, is to investigate how best we might translate some of the artistic framings of music – and the ways in which we think about creative Electroacoustic Music practices – to the world of Film Sound, so as to highlight the creative core at the heart of both. In so doing we hope to demonstrate that this practice has always been an artform, it was just that our ways of thinking and talking about sound have served to hide its artistic nature.

1.4.3 More than Technical – Towards More Diverse Fields

When we focus on the creative and aesthetic underpinnings of practice we shift the focal point of discussion toward that of communication and interpretation, de-centring the tools and the technology in favour of a more material and phenomenological approach. Technological discussions have value, but we should not limit ourselves to only ever engage with a subject via one perspective. By explicitly shifting the conversation we open up the potential for greater discussion of creative ideas and we may make these parallel practices more accessible to new and more diverse creatives who might otherwise be excluded by a discourse dominated by technology.

Evidence of structural access issues, within Film Sound and Electroacoustic Music is demonstrated by the symptoms of limited diversity representation. Both fields are dominated by white men (Butt 2020; Follows 2019), with only 5% of sound professionals identifying as women (Women's Audio Mission 2022). *Research from University of Southern California (USC) Annenberg highlights that this same disparity exists right across the sound and music industry, with a review of credits for 800 songs identifying only 5% of producers in the recording studio as female (Stacy et al. 2020).* And while it can be argued that there might no longer be formal structural exclusion, research has demonstrated that implicit biases and informal forms of exclusion are still very present.

The role of technological framings in music are explored by Georgina Born and Kyle Devine who reveal a stark divide in gender balance between Music and Music Technology subjects in UK higher education. Traditional music subjects (naively considered as more artistic) have an almost balanced gender divide of 55% female to 45% male, while music technology subjects are incredibly unbalanced with an enrolment of 10% female with 90% male. Born and Devine discuss how, "cumulative insights from feminist science and technology studies and the sociology of music education suggest that while girls and women are no longer formally excluded from scientific and (music-)technological pursuits, they are subject to observable processes of gendered exclusion – occupationally, discursively, spatially, and practically" (Born and Devine 2015, p.150). Substantiating this, key findings from the *Association for Electronic Music (AFEM) Gender Diversity in the Electronic Music Industry Report* highlighted that less than half of women who identify as a minority believe that their work is respected and taken seriously and that over 70% of women felt like that had to 'act like one of the boys' in order to fit in at work (Warren 2021).

Male (and usually white) dominance of the field acts as a self-fulfilling spiral maintaining its own exclusivity. While a focus upon technology feeds upon these cultural bias', further promoting overtly male engagement However,

> since the fabric of music production and engineering have been historically taken care of almost exclusively by males, there are reasons to presume that modes of listening, methods of production, approaches to creativity, mechanisms of priorities and so on are framed by a general hegemonic masculinity.
>
> (Marstal 2020, p.130)

Therefore, any way in which we can help to reframe narratives away from the technological has the potential to aid in the weakening of this dominance and the opening of alternative perspectives. This is not to deny the male standpoint or the technological. It is to simply recognise that these should not be regarded as the *only* ways to think about sonic practices. When curating contributions to this book, we deliberately sought to ensure a diversity of representation and to prioritise the voices of those who may have previously been marginalised or less heard. We are delighted that of the 17 contributors to this book, nine are women, including true pioneers such as Ann Kroeber, Annette Vande Gorne, and Jesse Dodd who is perhaps the only Black woman working in ADR and Foley Mixing globally. Through celebrating these practitioners we hope to reveal alternative ways of imagining and doing sound practice, the celebration of which has the potential to engage and attract ever greater numbers of diverse future practitioners.

1.5 What Did We Do, and Why?

Across autumn 2021 and spring 2022, we undertook 26 hours of semi-structured interviews with our 17 participants in order to access the experiences and knowledges of a diversity of specialisms and practices across Film Sound and Electroacoustic Music.

We conducted both individual interviews and set up dialogues between practitioners.[17] Drawing inspiration from Tim Ingold's work *Making* (2013), we sought to access the embodied, tactile, and tacit knowledges of our practitioners. While academic Electroacoustic Music composers are more familiar with sharing notions of their practice, they often do so in ways framed by the university system, which does not always encourage direct honest discussion of tactile embodied practices. We wanted to discuss *why* we create with sound the way we do, not to simply describe the individual steps in the signal path that modulate an audio signal, but to explore the creative impetus that drives us to structure and modulate sound and how this is tied to our intentions in communicating with an audience via sound, to transform impressions, spaces, emotions, and meanings through sonic intervention.

Our goal was not to seek one unifying rule that might underpin practices, but to identify a range of individual approaches and perspectives which guide and drive the creative processes of diverse participants. As noted by Donna Haraway, the situated knowledge of embodied subjects is necessarily partial and individually specific. While such situated knowledges cannot simply be added together, Haraway argues that shared conversations are possible but require the negotiation of complex networks of connection to identify meeting points within and around the differences that persist (Haraway 1991). Indeed, by bringing together a range of different perspectives and practices in our study we encourage a plurality which is aptly able to include the vistas of subjugated, marginalised groups – e.g. women – due to its core focus upon individuality and the adoption of difference as a core tenet. Embracing situated knowledges enables us to piece together insights in ways that do not need to conform to the forms 'domination, and the illusion of objectivity' so essential to conventional hegemonies of knowledge (Haraway 1989, p.222).

It was therefore a core goal of each interview to open up and reveal the individuality of each participant, to discover their unique approach and insight. As such, it was vitally important to the success of the research that both authors/interviewers were also practitioners working with sound. Our experiences as composers of Electroacoustic Music and sound design for independent and art film, provided us with our own tangible understandings of the material practices of working with sound, which enabled us to clearly develop a unique rapport with our participants via a shared understanding of the processes involved, and what it means to work with the medium.

While discussion of specific examples was useful in articulating key points in the dialogues, we tended to guide our conversations away from discussion of specific individual projects, and towards more underlying creative drivers. This delicate balance between specificity and generality was not intended to seek absolute rules or truths about practice, but to articulate underlying and consistent motives applied in the development and creation of projects. The subsequent dialogues were carefully curated to bring together practitioners who shared sensibilities or specialisms that we felt might speak well to one another and lead to compelling exchanges. In order not to influence or bias these discussions with prior contexts, we asked participants to share examples of their work so that there was material familiarity between the participants before speaking. These shared sounds were used as an impetus for dialogue and served as the ideal starting points for the conversations.

All interviews were transcribed and edited to ensure clarity for the reader. The conversations were co-collated to ensure maximum clarity with a balance of relevant detail, and each transcript was shared with the relevant contributor for final approval before publication.

1.6 With Whom Did We Talk?

We have been honoured to receive generous contributions from many leading professionals who kindly provided time within their busy schedules to talk with us. This book features contributions from nine Post-Production Film Sound professionals across ADR and Foley mixing, sound editing, sound design, sound effects, field recording, and re-recording (/dubbing) mixing; and eight Electroacoustic Music Composers working across the musical fields of acousmatic, soundscape, ambient, and sound arts.

1.6.1 Film Sound Interviewees

- Ann Kroeber – [Sound Designer, Effects Editor, and Recordist]
    - Renowned sound effects recordist, best known for her pioneering work with the Flat Response Audio Pickup (FRAP contact microphone) and her partnership with Alan Splet.
- Paula Fairfield – [Sound Designer and Sound Artist]
    - Multi Emmy award winning Sound Designer known for her iconic creature sound design featured in *Lord of the Rings and Game of Thrones.*
- Randy Thom – [Supervising Sound Editor, Sound Designer, Re-Recording Mixer]
    - Two-time Oscar winning Sound Designer and Director of Sound Design for Skywalker Sound.
- Nina Hartstone – [Supervising Sound Editor, specialising in Dialogue and ADR]
    - First European woman to win an Oscar for sound editing, she specialises in working with the intricacies of the voice through dialogue editing and ADR.
- George Vlad – [Sound Designer, Sound Recordist, and Composer]
    - Location sound recordist and Games Sound Designer, known for his capturing of rare and hard-to-reach soundscapes.
- Steve Fanagan – [Sound Designer, Supervising Sound Editor, and Re-Recording Mixer]
    - A regular collaborator with Lenny Abrahamson, he has received prestigious nominations for awards including wins from Irish Film and Television Academy (IFTA), an MPSE Golden Reel Award and an International Music+Sound Award.

- Jesse Dodd – [ADR and Foley Re-Recording Mixer]

  - With over 25 years' experience working at some of the world's leading studios she is one of the only Female African American Post-Production Sound Department ADR/ Foley Re-Recording mixers in the industry.

- Onnalee Blank – [Re-Recording Mixer and Sound Supervisor]

  - Multi-award winning Sound supervisor and Re-Recording mixer, known for her work and collaborations with Barry Jenkins and on *Game of Thrones*.

- Peter Albrechtsen – [Sound Designer, Re-Recording Mixer, and Music Supervisor]

  - Respected for his work across both documentaries and feature films, he works across a mixture of both Hollywood and Danish cinema productions.

1.6.2 *Electroacoustic Music Interviewees*

- Hildegard Westerkamp – [Composer, Radio Artist, and Sound Ecologist]

  - Her works explore acoustic ecology utilising recorded sounds of the environment and human voice. Instrumental in the founding of the *World Forum for Acoustic Ecology*, she served as chief editor of the journal Soundscape from 2000 to 2012.

- Annette Vande Gorne – [Electroacoustic Composer]

  - Founder of the *Métamorphoses d'Orphée* studio in Ohain, curator of the festival *L'Espace du Son* in Brussels. She studied with Guy Reibel and Pierre Schaeffer at the *Conservatoire national supérieur* in Paris and worked with François Bayle at the GRM.

- John Young – [Electroacoustic Composer]

  - Award winning New Zealand Composer, his works merge sound-images of the real world with more abstract sonic materials, inviting listeners into imaginative worlds where familiar objects and environments are given new meanings.

- Trevor Wishart – [Composer and Free Vocal Improvisor]

  - A Composer and free-improvising vocal performer, working primarily with the human voice. Recognised for both his composition and extensive writings on the topics of sonic art, composition with sound, and the morphology and psychoacoustics of sound transformations.

- KMRU – [Sound Artist, Experimental Ambient Musician, and Producer]

  - From Nairobi, his works deal with discourses of field recording, improvisation, noise, ambient, radio art, and expansive hypnotic drones.

- Annie Mahtani – [Electroacoustic Composer, Sound Artist, and Performer]

  - Has worked extensively with dance, theatre and on site-specific installations, she has a strong interest in field recording, and exploring environmental sound.

- Nikos Stavropoulos – [Composer of Acousmatic & Mixed Music (Instrument and Electronics)]

   – His practice is concerned with notions of tangibility and immersivity in acousmatic experiences and the articulation of acoustic space, in the pursuit of probable aural impossibilities.

- Natasha Barrett – [Composer]

   – She explores new technologies and experimental approaches to sound in a broad range of contemporary music, including concert works, public space sound-art installations, and multimedia interactive music.

1.6.3 List of Dialogues

- Ann Kroeber with Hildegard Westerkamp

   – Two female pioneers of sound practice with a shared sensitivity to listening to the environment and the communicative potential in non-human animals and insects.

- Nina Hartstone with Trevor Wishart

   – Leading experts on the expressive potential of the voice and its articulations within and beyond language as a powerful communicative tool.

- George Vlad with KMRU

   – A discussion built around a shared sensitivity to place and location and the identities of sounds within an environment.

- Steve Fanagan with Annie Mahtani

   – Two artists both interested in constructing with impressions of space and spatiality discuss their approaches to developing deep and multilayered sonic worlds.

- Onnalee Blank with Nikos Stavropoulos

   – A shared interest in the performative nature of gesture, and the articulation of spaces and layering for creative effect.

1.7 Summary

The priority of this research has been to highlight and better understand processes of creativity in sound practice across both Film Sound and within Electroacoustic Music. We hope that by exploring a broad range of practices, we can contribute to their demystification, enabling a greater appreciation of their artistry and the diverse potentials of sound as a creative medium.

   In this opening chapter, we have set out our aims and objectives and provided some of the contexts against which we have situated our research. In the next chapter, we will discuss in more detail our approach and methods as well as some of the rationales behind them.

Notes

1 Descriptive or declarative knowledge is constituted by facts, 'knowing-that'. It is contrasted with procedural knowledge which is about process, 'knowing-how'.

2 Our research has been funded by the UK Arts and Humanities Research Council (AHRC) via a Leadership Fellowship grant "Audiovisual Space: Recontextualising Sound Image Media" – For more information visit: https://www.gre.ac.uk/research/groups/sound-image.

3 Due to practical limitations in the scope of the research, it was not possible to include production sound professionals as part of this study. But, there is strong potential for a new and additional study to bring together leading music recording engineers and professionals with production sound mixers to explore potential synergies in more explicit sound capture and recording contexts.

4 Even lines of dialogue from principal actors can be re-recorded and replaced in Post-Production through Automatic Dialogue Replacement (ADR) recording sessions in the studio.

5 An overview of the history and main creative movements that have led to the development of contemporary Electroacoustic Music can be found in the chapter *Electroacoustic Music: An Art of Sound* (Knight-Hill 2020b) within the Routledge Textbook *Foundations in Sound Design for Linear Media* (edited by M. Filimowicz).

6 A technique which ensures that all 12 notes of the chromatic scale are sounded as often as one another to provide equal importance of all 12 pitch classes.

7 For example, GRM Tools plug-ins created by Group de Recherches Musicales a pioneering French sonic research centre in electroacoustic research founded by Pierre Schaeffer in 1958 as part of Radio France. https://inagrm.com/en.

8 An array of loudspeakers are placed everywhere across a concert or event space, surrounding the audience.

9 Heightened listening is a focused and attentive approach to sounds in which the listener focuses upon the details of sounds, perhaps moving away from source associations.

10 Tim Ingold is an anthropologist. His research questions how greater appreciation for environmental perception and skilled practice can provide alternative perspectives to Western scientific hegemony. (He also plays the 'cello). See also:

   Ingold, T. (2000) *The Perception of the Environment: Essays on Livelihood, Dwelling and Skill*. London: Routledge.
   Ingold, T. (2007) 'Against soundscape'. In A. Carlyle (ed.) *Autumn Leaves: Sound and the Environment in Artistic Practice*, pp.10–13. Paris: Double Entendre & CRISAP.
   Ingold, T. (2013) *Making: Anthropology, Archaeology, Art and Architecture*. London: Routledge.

11 Georgina Born is an anthropologist and musician. Her research has critiqued sites of institutionalised cultural production with notable studies on the musical research centre IRCAM (Paris) and the BBC and the transformation of musical practices by the shift into the digital. See also:

   Born, G. (1995) *Rationalizing Culture: IRCAM, Boulez, and the Institutionalization of the Musical Avant-Garde*. Berkeley: University of California Press.
   Born, G. (2005) *Uncertain Vision: Birt, Dyke and the Reinvention of the BBC*. New York: Vintage.
   Born, G., & Hesmondhalgh, D. (eds.) (2000) *Western Music and Its Others: Difference, Representation, and Appropriation in Music*. Berkeley: University of California Press.

12 Donna Haraway is a philosopher of science best known for her feminist critiques of gender bias in scientific research and her writings on humans and technology and culture. See also:

   Haraway, D. (1989) *Primate Visions: Gender, Race, and Nature in the World of Modern Science*. New York: Routledge.
   Haraway, D. (1991) *Simians, Cyborgs and Women: The Reinvention of Nature*. New York: Routledge.

13 Music was one of the elements of the *Quadrivium*, a grouping of four arts subjects (arithmetic, geometry, astronomy, music) that formed the second-stage of the ideal Liberal Arts education. First defined by the ancient Greek philosopher Plato in his book *The Republic*, this four-part structure was intended to build upon students first-stage studies of the *Trivium* (grammar, logic, and rhetoric). And these two-tiers of seven key subjects became the foundation of the Medieval University.

14 The recording of sound onto magnetic tape had been perfected by German engineers in the early 1930s but the technology was not widely available outside of Germany until a series of Magnetophon recorders were taken back to the USA and UK at the end of the Second World War.

15 "[T]here remains a strong foothold of the aesthetic which banishes real-world sound-identities as 'non-musical'" (O'Callaghan 2011, p.55).

16 As noted by the Sound Credit Initiative in their open letter, "Production Sound Mixing, Sound Editing, and Re-recording Mixing [...] are generally not afforded prominent screen credits that are representative of their creative contribution to the film" (Sound Credit Initiative 2020).

17 Inspired by the iconic exchanges between Karlheinz Stockhausen and 1990s emerging electronic musicians, as engineered by Richard Witts and documented in his iconic article *Stockhausen Versus the Technocrats* (2004).

2   Making Sense of Making

Our Approach

In this chapter, we discuss our practical approaches to undertaking this research and why we decided to use these methods. In setting out our methodology, we hope to promote the ethos underpinning our research and encourage and inspire future researchers engage in their own studies of creative practice, whether that be in sound or in associated creative disciplines.

2.1 Embracing Complexity

Making is complex; drawing upon experience of prior practice, discussions with fellow creatives, demands of work schemes and the affordances of the materials and tools in play. This complexity can be even more challenging to explain to others, especially when many facets of the process can be intuitive, tacit, and perhaps unconsciously enacted. Our role as researchers has therefore been to draw out these implicit details from leading professionals, to ask questions that encourage conversations which elaborate and elucidate insight into what really takes place within the processes of making, so as to provide greater access to underlying principles, approaches and techniques at play.

Every project in sound establishes a unique set of requirements, boundaries, and potential for creativity, and each practitioner will respond to those demands in their own way. It is therefore reductive and limiting to imagine that there might be a single 'ideal' answer to the questions of "What makes great sound design?", "How do you compose the perfect sound gesture?", or "What constitutes effective sound practice in music or film?".

There are as many answers to these questions as there are new projects, multiplied by the number of creatives who might work on them. Thus, to look for universal 'rules' in this diverse potential of possibility is to miss the point. As a result, our goal has never been to seek a singular answer, but to highlight the many possibilities that exist within the rich interrelated web of inspirations and actions, which can be drawn upon to inform, drive, and shape practices of sonic creativity.[1] It has been our mission to find a solution that enables us to embrace the inherent messiness of sonic creativity and treat this complexity, not as a problem which needs solving, but as a vital function of the very process itself.

DOI: 10.4324/9781003163077-2

2.2 Foregrounding Process

Our project has borrowed much from the discipline of anthropology to help frame and guide our research, but we are not anthropologists by training. As researchers we are primarily composers and practitioners – we make with sound. As such, it is a fundamental curiosity about how we might go about making that drives us to find out how others approach and enact their making: "Do others create in the same way?", "Are there any shared techniques?", and "How can we best discuss our practices to share approaches and further our understanding of one another?".

Our own experiences of making[2] provided us with knowledge about *doing* practice, which has been invaluable in guiding us to unlock insights about the practices of others (even if the practical implementations are very different to our own). As composers experienced in working with sound, we were in some ways able to capitalise on a shift from a positionality as authoritative external observers, to something more akin to 'indigenous ethnographers' (Fahim 1982; Ohnuki-Tierney 1984). This approach helped to shift the balance of power within our discussions, away from us being externalised academic observers, towards relationships of greater collegiality. Essentially, the fact that we are all 'workers with sound' enabled us to engage in informed, empowering and genuinely collaborative discussions about creative processes.

Studies of Electroacoustic Music and Film Sound can often focus on specific individual sound objects (e.g. what technique did Ben Burtt[3] use to create the sound of the lightsaber[4] in Star Wars?[5]) or specific material details of recent projects in isolation (e.g. "tell us where the sounds in your latest electroacoustic composition came from"), and while these types of enquiry can seem essential on the surface, they lend themselves towards oversimplification. A description of what was recorded, or the editing steps taken to create a specific sound object, can provide interesting snippets of information and are easy to document, but they belie the underlying creative truths that hide beneath them. Knowledge of *what* was done tells us little about the ways in which creativity operates or how these aesthetic decisions were reached. In fact, valuing the descriptive and immediately qualifiable often masks the possibility of delving into deeper questions around processes of making and the underlying aesthetic drivers of practice, by closing off questions around processes of choice and creative possibility. In such situations we fix sounds and can come to believe that the final result is a *fait accompli*, an absolute fact that was never in doubt. This actively masks that the final outcome is developed in an unfolding process, and emerges as one out of many possible choices. As Ingold notes:

> In the study of material culture, the overwhelming focus has been on finished objects and on what happens as they become caught up in the life histories and social interactions of the people who use, consume or treasure them.
>
> In the study of visual culture, the focus has been on the relations between objects, images and their interpretations. What is lost, in both fields of study, is the creativity of the productive processes that bring the artefacts themselves into being: on the one hand in the generative currents of the materials of which they are made; on the other in the sensory awareness of practitioners.
>
> Thus, processes of making appear swallowed up in objects made.
>
> (Ingold 2013, p.7)

To fixate on final objects as the sole locus of knowledge is to deny deeper understandings of practice. Further, to retain a focus on the declarative "what did they do?" can inadvertently perpetuate the idea that sound practice is a purely technical craft with little creative depth, placing attention on the mechanical routines of realisation rather than the deeper creative thought processes underlying them. If we seek to understand things in direct or simplistic terms, then we naturally end up with simplistic explanations. If we seek to understand things in direct or simplistic terms, then we naturally end up with simplistic explanations. As such, practitioners can inadvertently mask their own creativity through overly descriptive expositions of their craft.

By directing our focus away from objects made, towards creative possibilities and the choices available within processes of creation, we have focussed attention on the process of making and *why* decisions are taken. This approach is more able to reflect the complexity of soinc practice and the many possible options open to each creative. As a result, we believe that we have been able to access new routes of conversation that provide richer perspectives upon creative practice in Film Sound and Electroacoustic Music.

2.3 A Fusion of Many Voices

Every field constructs its own frames of reference and terms of analysis, and thus conventions develop which re-enforce those assumptions and designs. These conventions can become so pervasive that they begin to appear self-evident from within and act to shape how the field defines itself[6]. Their ubiquity can make them appear as fundamental 'facts': objective truths that cannot be questioned.[7] We have already discussed Ingold's contrasting forms of knowledge; either knowing *about* something (what was done) or knowing *through* (why it might have been approached in this way) in the previous chapter and how these framings applied in discussions of sonic creativity can have significant implications of how we access and understand practice. But an additional challenge for the fields of both Film Sound and Electroacoustic music is that many prior dialogues with practitioners take the form of individual isolated discussions which often focus upon one practitioner and their approach to one particular project. Thus, because these perspectives are isolated examples bound by specific individual contexts, it is impossible (or unreliable) to draw more broad findings or conclusions from them. If we are to access more reliable truths about practice, we needed to find ways in which we might liberate ourselves from both traditional conventions and isolated perspectives.

Our strategy to overcome the conventional framings of knowledge has been to layer multiple perspectives. Our hypothesis was that through bringing practitioners together from the parallel worlds of Film Sound and Electroacoustic Music we might disrupt the conventions of each, and in so doing find a way to access any underlying creative principles, unlimited by the standard frames of reference for each field. Thus, though our project is far from the first to engage in interviews with practitioners, we have been fortunate to be able to engage in a focused study with many practitioners simultaneously, applying a coherent methodological frame and approach to elicit multiple perspectives. This multiplicity of voices has enabled richer analysis and afforded new insights.

Many of the voices in existing literature are often from a narrow demographic (white western males), and there is a risk with such a lack of diversity that key insights from alternative perspectives are lost. As noted in the last chapter, in order to attempt to reflect a diversity of approaches to creativity, we actively sought out contributors across a wide range of roles, specialisms, and creative practices from within each of the fields of Film Sound and Electroacoustic Music, and we sought to assemble a group of participants which offered more inclusive representation to those voices which might not always so well heard in other texts and publications.

An underlying ethos of this project has been to bring together creatives to discuss their practices and together seek new insights. We were aware that some of the most compelling outcomes might be revealed in differences and disagreements between the fields of Film Sound and Electroacoustic Music, therefore we curated specific dialogue pairings informed by potential similarities between the practitioners, which might also, in their proximity, reveal compelling tensions for example, Nina Hartstone and Trevor Wishart who both work extensively with the human voice; or George Vlad and KMRU who both work with field recordings and have a keen engagement with place in their work.

2.4 Doing Conversation

Our primary approach to data collection was the conversational interview. We used the 'semi-structured interview' format to guide our conversations across both the individual interviews and the dialogues. In this, a schedule of key topics provides a basic structure of discussion points, from which conversations could unfold fluidly in their own direction. Our schedule needed to include issues that would trigger compelling discussions, but without being too prescriptive.[8] The schedule developed out of our own prior research (see: Hill 2013, 2017; Knight-Hill 2019, 2020a; Margetson 2021) which critiques how creative sonic practice is framed, imagined, and contextualised. These projects have been increasingly inspired by notions of the 'spatial turn' in the humanities and how this movement might pertain to music and sound arts (see, Knight-Hill 2020a). However, it became quickly clear that explicitly introducing philosophical arguments into conversations was highly disruptive to the flow of exchanges with practitioners, and therefore these theoretical underpinnings were most beneficial when they remained beneath the surface, informing and guiding the direction of questioning but without imposing themselves as abstract distractions to the conversations in hand.

We did not always engage each practitioner with every single item on our schedule of topics, choosing instead to value depth and the richness of emerging ideas, rather than full coverage of our pre-conceived topics. In addition, we were ready to reflect and conduct analysis in situ, presenting back our own interpretations and summaries of concepts we thought were being advanced to the interviewees for cycles of clarification and deeper discussion. Because, we were seeking emergent outcomes from genuine conversations, we needed to be open to digression, tangents, and the unstructured dialogic interaction of an open conversation. We also needed to ensure that dialogues were able to represent the diverse individual conditions of each practitioner's specialist practice and would not artificially limit the scope of our discussion as a result of a priori assumptions. By engaging the subjects as active participants within a collaborative co-production of knowledge, our open conversations had the potential to reveal findings that fell outside of our own original schedules of questioning. This open process enabled us to access new avenues and topics of discussion in which the flow of such exchanges often enabled deeper access to the complex realities of creative practice than would ever have been elicited by our pre-conceived schedule of key topics alone.[9]

To prepare for our interviews and discussions we additionally engaged in contextual preparation via thematic research into individual practitioner's discourses. This included exploring any previous writings, prior interviews, past work and, in some cases, their contributions to online discussion fora. These readings helped contextualise and fill out our understandings of responses elicited within our conversations and enabled us to develop a complementary set of curated individual questions for each practitioner based upon their individual specialisms and oeuvre.

2.5 Dialogues and Exchange

Our dialogues – in which we brought together a Film Sound practitioner and a Composer of Electroacoustic Music – was initially envisaged as being the apex of the methodology, the moment of most valuable interaction, but primarily they often provided points of triangulation; fresh insights upon previously cited information. When similar examples or concepts were discussed between the solo interviews and pair discussions, their position in the context of a new conversation always revealed new subtleties and detail. Sometimes this involved delving deeper into specifics, or it might have revealed more tangible practicalities from a new perspective. Items repeated were clearly of greater value to the participants and thus, with all of this information across multiple conversations, we were able to build up a more richly informed picture of how each participant approaches and values their own creative process.

A secondary impact of introducing two creative practitioners in dialogue was that we shifted the researcher dynamic of our exchanges and created a new context within which we (as researchers) were able to sit outside the direct conversation, more as observers reflecting upon the similarities and differences, to both triangulate our findings from individual interviews and derive new insights.[10] Our role in the dialogues was intended to be as minimal as possible, a facilitator of discussion only lightly intervening to guide the trajectory on occasion. Prior to the dialogues we kept the identity of the pairings secret, instead seeking to facilitate initial introductions via listening examples.[11]

We asked each participant to share excerpts of their practice, up to ten minutes in duration, which we passed on anonymously to their partner prior to the dialogues. In this way we centred the sonic as the initial jumping off point for the dialogues, providing a sound based introduction as a route into one another's working practices. This helped to demonstrate a basic level of commonality and provide a jumping off point for the conversations lessening the potential alienisation of the exchange, which otherwise might have biased or disrupted free and open conversation.

We asked practitioners to listen to these examples in advance of the dialogues and to record any notes in response to their listening. Details of the specific sound examples discussed can be found within the introductory preamble to each pairing.

One of the most affirming aspects of the dialogues was the realisation in the participants of similarities between their practices, aesthetics, and approaches across Film Sound and Electroacoustic Music. Each exchange reflected a palpable sense of collegiality and inquisitive engagement, sharing material processes and examples, demonstrating the affordances interchange between these contiguous fields. These dialogues were fundamental to the collaborative ethos of the project, bringing together not just ideas and concepts, but people to discuss their material processes directly.

2.6 A Collaborative Process

In the development of robust texts for analysis and publication in this volume, we utilised a co-production approach. We provided the practitioners with an edited transcript and invited them to make changes and suggestions. In this way, each practitioner became an equal contributor in the formation of their own expression, which allowed free flowing discussion in spoken conversation to be clarified and checked to ensure accurate articulation of specific details. This process of collaborative back and forth between researchers and contributors is what the musical anthropologist Steven Feld calls 'dialogic editing' (Feld 1987), a process of co-editing, which breaks down the hierarchy of 'researcher to subject' in order to directly engage the contributors themselves within the process of their own representation.

In conversation, each participant often introduced multiple examples (e.g. compositions, sound design in film) to describe and explain similar points within their interview. Within a transcribed text this can quickly become difficult to read and the complexity can mask the underlying point. As such, we worked with each contributor to select the most concise and relevant examples for each of their points, so as to highlight the key underlying message that they wanted to convey.

As we engaged more extensively with the texts, we began to identify more comprehensive themes which we annotated into the texts with subtitles. These subtitles aided our comparison across and between interviews, and eventually evolved into the more accessible section subtitles that you find in the interviews and dialogues. These subtitles are not directly consistent across all interviews because we were not seeking to apply a set of strict rules to bend what people say into our own academic categories. But, we found that some form of subdivision was useful in framing specific topic discussions and segmenting the text for readability and ease of navigation. Where there has been a clear synergy across interviews and dialogues we have used the same subtitles, in a limited fashion.

We did not set out with a preconceived analytical structure into which we hoped to fit our findings.[12] Instead, we made continual comparison, reflecting back on prior interviews in a cyclical fashion, enabling threads and similarities to emerge and the main findings to reveal themselves. Therefore, our process of analysis and making sense of individual contributions was open and procedural, constantly evolving.

2.7 Summary

In this chapter, we have discussed the challenges of documenting diverse sonic practices and the practical steps taken to ensure rich and compelling insights. We have set out our approach so as to provide a context for the conversations which appear within this book and to provide potential inspiration to other researchers who might embark upon similar studies. If we accept that the realities of practice are complex then we need to adopt approaches that can embrace and celebrate that complexity; engaging with diverse perspectives and digging deeper into artistic processes.

By setting out our process here, we hope to demystify some of the potential approaches that can be applied and encourage others might engage in future research to capture the experiences of practitioners from other fields, other nationalities, and other ethnicities to further broaden and continue the exploration of sonic practices.[13]

As noted previously, our findings are necessarily limited by the people that we asked and the geographical and historical context within which they operate and live. Each contributor brings with them what James Clifford calls a 'partial truth' (1986), and we have sought through our curation of participants to provide a range of views and perspectives which reflect a broad scope of creative sonic practices that bring together many partial truths in a way that might reflect a wider range of the possible experiences and approaches available. We have been delighted to feature so many contributions from women, who have historically been marginalised, but we do recognise that the majority of contributors are white western practitioners based in the UK or USA.

In the next chapter, you will find the results of our analysis and reflections which evolved into Our Map of Sonic Creativity.

Notes

1 As Gilles Deleuze asserts in his discussion of sense informed knowledge "the aim is not to rediscover the eternal or universal, but to find the conditions under which something new is produced" (Deleuze and Parnet 1987: vii).
2 Knowing through.
3 Ben Burtt is a Sound Designer, Film Editor, Director, Screenwriter, and Voice Actor. He has worked as Sound Designer on various films, including the *Star Wars* and *Indiana Jones* film series, *E.T. the Extra-Terrestrial*, *WALL-E*, and *Star Trek*.
4 An energy sword used by characters in the *Star Wars* franchise.
5 *Star Wars: Episode IV A New Hope* (1977). Directed by George Lucas. [Film] US: 20th Century Fox.
6 "Every artist, intermediary and institution necessarily defines their position and that of the genres or artworks that they advocate in terms of relations of difference or opposition to others within the field, while the field as a whole is 'governed by a specific logic: competition for cultural legitimacy'" (Bordieu in Born 2010).
7 Jonathan Sterne's critique of the 'Audiovisual Litany' is a deconstruction of how many of our ideas about sound are based in assumptions about Judeo-Christian ontology, derivative of ideas of sonic salvation (Sterne 2003, p.15).
8 Our key schedule of topics included the following:

- Do you have a personal aesthetic style? – use of tropes, etc.
- How do you know when you have discovered the 'right' sound?
- What significance do you place upon perceived reality when working?

    - Real/unreal and its position in constructed narratives.

- Do you think in terms of materiality & gesture (texture/gesture)?
- Is your creative process pre-planned or emergent?

    - Are you identifying specific points and then weaving them together, or creating an overarching atmosphere within which you then choose to articulate specific points?

- To what extent are you using sound to articulate or convey feeling?
- Audience – how do you know something will communicate or connect with an audience?

9 We are grateful to Prof. Georgina Born who acted as a mentor for us in anthropological research methods, providing bespoke tutoring, and informing our research methods and approach with her extensive experience and expertise.
10 "[E]thnographic data depend on the social relationship of researcher with subjects, research reports must clearly identify the researcher's role and status within the group investigated" (LeCompte and Goetz 1982, p.38).
11 The only time that the identity of a partner was revealed in advance was with Hildegard Westerkamp as we needed to provide an additional rationale to convince her of the value of

participating in the study. As such, we shared our plans to pair her with Ann Kroeber in advance of their dialogue.

12 "Rather than devising research protocols that will purify the data in advance of analysis, the anthropologist embarks on a participatory excursive which yields materials for which analytical protocols are often derived after the fact" (Strathern 2004, p.5).

13 We also encourage readers to investigate the work of our colleagues Dr Amit D. Patel and Dr Gabrielle Messeder, 'Disruptive Frequencies: Black and South Asian Sound Artist in their Own Words' (forthcoming 2024) which explores creative practices of underrepresented artists in the UK Experimental Music scene.

3 Mapping Stories of Creative Sonic Practice

As we engaged with the stories shared with us, we began to identify recurring trends and common approaches. By listening to and co-editing the interviews and dialogues, we reflected deeply upon the responses given and, via a slow process of distillation, gradually assembled and highlighted key creative decision making processes.

We have rendered the key themes into what we call 'Our Map of Sonic Creativity'. The goal of such a map is not to seek to categorise or log abstract facts, but to draw attention to and appreciate the true depth and richness of creativity that is deployed in the articulation of sound. Some entries may be obvious and direct, while others embrace more tacit, implicit, unconscious, and intuitive considerations and, as a result, reflect approaches which might be enacted in an unconscious way. At the most fundamental level, each point on our map reflects an opportunity for creative choice within the decision-making process. In highlighting these possible points of creative choice, we hope to make apparent the artistic opportunities available for creative input, and to affirm the agency of the creative in being able to consciously make the decision that best suits their own creative intention for the project at hand.

As we set about identifying key points we found that some insights were explicit – clearly expressed within our interviews and dialogues and directly comparable – while others were implicit – emerging only after we had the time to reflect, digest, and understand the underlying process and its implications.[1] In many cases we used a process of triangulation to cluster comments and associations which referred to similar common creative factors, but which were themselves unique. As such, we identified emergent points of synergy which might not have been apparent if we had interviewed fewer participants, or not had a mixture of Film Sound and Electroacoustic Music practitioners to talk with. Thus, this is not a comparative study that isolates and objectifies difference between practices or practitioners, but one that seeks to draw connections and reveal commonalities.

Many of the contributors to this volume mentioned how rare and valuable it was to engage in considered reflection and deep discussion about processes of making. We hope that this book is just the start of such ongoing exchanges as each reader reflects, evaluates, and appreciates the richness of their own sonic practice and recognises more fully their own opportunities to drive their personal creative aesthetic.

DOI: 10.4324/9781003163077-3

3.1 Our Map of Sonic Creativity

To ask for a map is to say, tell me a story.

(Turchi 2004, p.11)

Ours is a story about the processes of creative decision making, inspiration, and doing. Our map seeks to identify the markers, waypoints, and topography of potential soundscapes; offering you the opportunity to chart your own pathways into creative soundworlds.

As noted above, our map does not provide an explicit set of instructions for *what you should do*. Rather, it sets out waypoints and avenues of possibility, enabling you to navigate *how you might do*. Readers may instantly recognise themselves in the discussions of practices below, or they may discover new categories via their own readings of the interviews. In providing this map, we hope to enable *new* opportunities for reflection, opening our acoustic horizons to celebrate the detail, and care of the processes that we apply in making. Perhaps offering an opportunity to reflect upon and consider our own approaches to creativity and decision making, while supplying a vocabulary of ideas and approaches that enable discussion with others.

Practitioners may choose to apply as many or as few of the following *Principles, Approaches,* and *Techniques* in their creative workflows, varying their application across, between, or within projects. And this is why a map is the perfect metaphor. It is not a linear series of discrete steps or an explicit list of serial instructions, it is a palette of potential.

Each entry across our map emerged from our reflections upon the conversations collected within this volume, drawing from the experiences of 17 world leading professionals to identify three main categories:

- PRINCIPLES – Foundational ideas that inform a core philosophy of making.
- APPROACHES – Aesthetic strategies deployed to articulate sound and evoke affect.
- TECHNIQUES – Material practices through which ideas and approaches can be articulated in sound.

These items have been represented in graphical form, arranged in concentric circles, and colour coded,[2] but we shouldn't think of our map as existing only in the form of its graphic representation on the page, but as a dynamic constellation of ideas. Thus, the detailed definitions of individual terms below also constitute a map, but in a different form. You might use the graphical map to identify key points of interest and then look up their detailed definitions in the chapter below, or seek out the detailed definitions and then reflect on their orientation and relationship to other concepts by visiting the graphical map representation.

Our goal has been to help reveal the many possibilities available, so that you can orientate yourself within the rich complexity available to you. The whole point of describing our ideas as a map, rather than a table, chart or list, is to welcome the flexible drawing out of correspondences. We have set out a terrain, shaping its topography through the significant points that emerged across our discussions, but this is not to claim that we have identified every possible point of significance. Indeed, we have been careful to position our map so that it does not operate as an 'absolute' or 'fixed' model of rules or instructions for describing how sound practice is done. Such an objectification of practices could not be further from our intention. As a result, our map is filled with terms that invite question, engagement, and reflection; and which welcome you to develop your own perspectives. Ours is a desire to open up and create dialogue, not to codify or lock down.

We envisage the map as a tool for thought and reflection, a space upon which you can annotate and draw your own trajectories. We have included some indicative dotted annotations within the graphical map to demonstrate potential links between points, but have been careful to avoid making connections which imply fixed links as it is important that the items are considered as non-hierarchical and independent.[3] You might choose to photocopy our graphical map from this book or download a digital colour version for printing as an eResource from the Routledge website [www.routledge.com/9780367755881].

3.1.1 How Might I Use This Map?

Our map might be used to navigate an existing sonic terrain or identify new opportunities to chart out your own soundscapes. Just as on any journey, you might approach our map in a range of different contexts and with varying goals in mind. In much the same way, there are many possible solutions to creative challenges, and rather than seeking the one 'perfect' or 'correct' answer, our map recognises and celebrates that there is never just one truth, but many possible truths.

You might choose to:

- Consider links between ideas and practices.
- Analyse and reflect upon creative processes (your own, or others).
- Identify possible novel extensions for existing decisions.
- Stimulate group discussion when working collaboratively.
- Deconstruct and consider possible creative intentions in an analysis of someone else's work.

You might choose to make use of our map at any stage within the creative process. Perhaps for analysis and reflection upon completed works or in the generation of new works. It might be used at the start of a project to generate ideas (either alone or in dialogue with collaborators), or part way through the project as a point of evaluation, reflecting upon the path in which the work has developed since its initial inception.

The main objective of creating our map has been to set out the terrain, so that you can clearly identify the range of opportunities open to you. Our hope is that with our map you can reflect upon which approach you might take, and ensure that you are not missing other available opportunities and to avoid falling into stale conventions and self-limiting your potential.

3.1.2 Definitions of Terms

Each term found on the graphical map is defined in detail within the chapter below. Detailed descriptions are set out page by page for each entry, listing a series of verbatim practitioner quotes, drawn directly from our conversations and dialogues, alongside each. Gathering multiple quotes together demonstrates two aspects:

1 A direct link to original conversations, a concrete demonstration of how each contributor referenced each item.
2 How the plurality of perspectives from multiple practitioner's provides a unique insight which enabled us to define each item.

In listing the multiplicity of approaches and intentions from varied practitioners alongside each map entry, we hope to have reflected applied examples and something of the multiplicitous reality of creative practice. The quotations help to elaborate the various directions in which these central concepts might be applied and demonstrate the multiplicity of perspectives which have led to our identification and triangulation of these points.

Together these constellations of quotations reveal more than any one dialogue could have revealed on its own. In some cases, our categories may stem from explicit statements made by a number of individual practitioners, but in many cases they are emergent topics, developed from the synergy and synthesis of our own reflection and interpretation of what our participants said.

3.1.3 *Three Dimensions* – Principles, Approaches, Techniques

Each of the three categories in our our map relates to a distinct modality of engagement with sound, and because of this it is useful to think about our map as more than just a two-dimensional surface, but as actually occupying a three-dimensional space. We've reflected this in the graphic design, making use of the curved topography of the physical page in this book, which positions each of the three categories at a different height.

Underlying *Principles* sit lowest in the centre (reflecting their role as core underlying ideologies of practice), with *Approaches* above and expanding out from these (as opportunities to apply and explore the articulation of sound). *Techniques* sit closest to the 'surface' as they reflect more directly tangible features of the soundscape. As such when we are drawing associations across the map, we are not just doing so in two dimensions but in three (Diagram One).

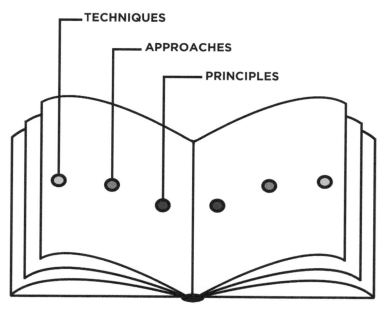

Diagram One Demonstrating the three-dimensional depth of the map in this book.

This means that, while we can't always draw direct equivalences between points, there will always be relationships between them. *Principles* are more abstract underlying approaches, while *Techniques* are directly tangible articulations of sound. Bridging the gap are *Approaches*, offering frameworks of application for both underlying principles and surface techniques.

3.1.4 A Note on Formatting

Where a quotation came from one of our interviews, the individual artist's name is listed:

e.g. **Hildegard Westerkamp** – "[quote]".

Where the quotations are drawn from one of the dialogues between practitioners we cite the speaker, while referencing their partner in conversation:

e.g. **Hildegard Westerkamp** (in conversation with Ann Kroeber) – "[quote]".

For ease of reading, and to focus upon the ideas, we have not included direct page number references for each quote. Quotes are reproduced here to directly exemplify and demonstrate the *Principles, Approaches*, and/or *Techniques* identified.

3.1.5 Examples of Possible Applications

Here are some examples of possible use cases for our map:

- **Generate Ideas** – You're seeking to orientate yourself without any fixed points of reference. Use the map to identify possible avenues for exploration.
- **Nucleus** – Begin with one point on the map and draw out relations to this via association.
- **Compare** – Select two or more points and consider their synergies and potential relationships.
- **Building Up** – You have already identified one or more key *Principles* that you want to elaborate in your work. Use the map to build out across *Approaches* towards practical *Techniques* that you might employ in realisation of this.
- **Digging Down** – You have identified one or more material *techniques* that you want to explore. Map inwards through relevant *Approaches* and towards underlying *Principles* that help underpin and fully elaborate your technique.
- **Dialogue** – Use the categories as points of discussion within collaborative exchanges to trigger conversations about ideas, seeking to access underlying creative imperatives for better collaboration and partnership.
- **Develop** – Use it as a teaching tool or for professional development activities. Explore each concept in turn, reflecting upon its application in a range of contexts and in relation to other concepts.

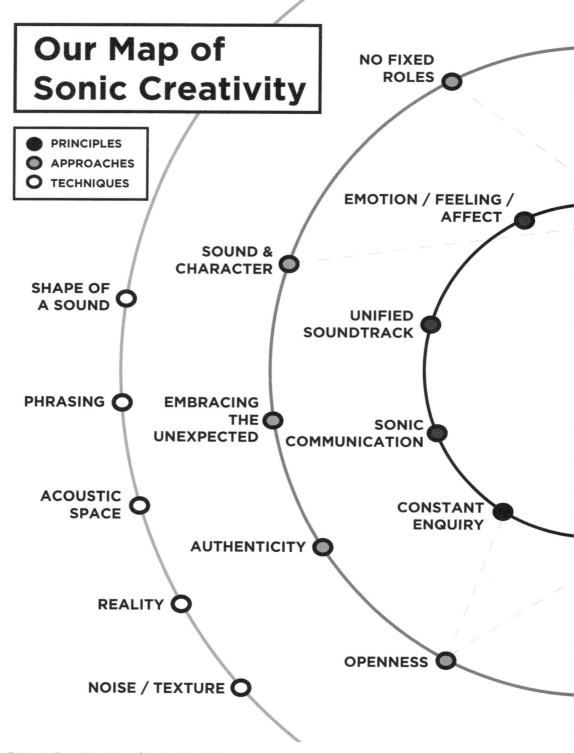

Our Map of Sonic Creativity

PRINCIPLES
APPROACHES
TECHNIQUES

NO FIXED ROLES

EMOTION / FEELING / AFFECT

SOUND & CHARACTER

SHAPE OF A SOUND

UNIFIED SOUNDTRACK

PHRASING

EMBRACING THE UNEXPECTED

SONIC COMMUNICATION

ACOUSTIC SPACE

AUTHENTICITY

CONSTANT ENQUIRY

REALITY

OPENNESS

NOISE / TEXTURE

Diagram Two Our Map of Sonic Creativity.

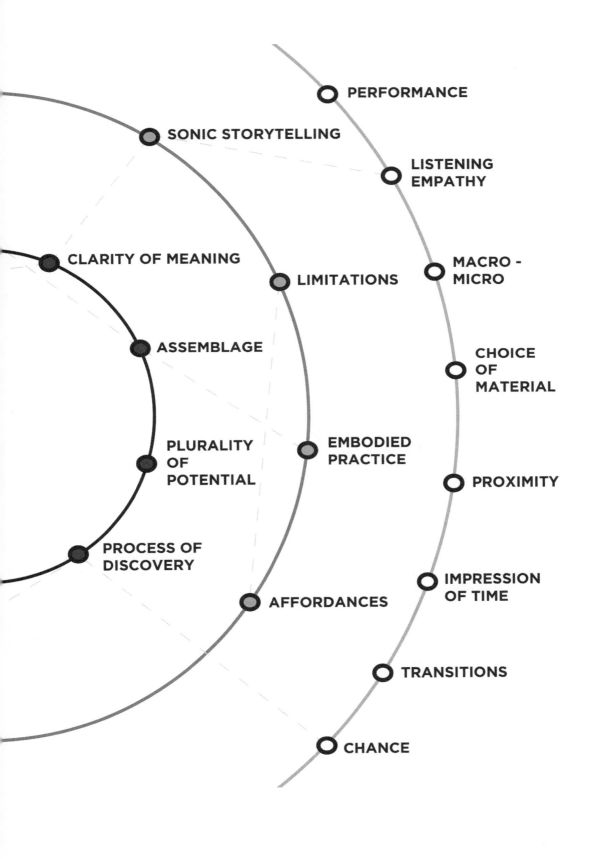

PERFORMANCE

SONIC STORYTELLING

LISTENING
EMPATHY

CLARITY OF MEANING

LIMITATIONS

MACRO -
MICRO

ASSEMBLAGE

CHOICE
OF
MATERIAL

PLURALITY
OF
POTENTIAL

EMBODIED
PRACTICE

PROXIMITY

PROCESS OF
DISCOVERY

AFFORDANCES

IMPRESSION
OF TIME

TRANSITIONS

CHANCE

3.2 Principles

Foundational ideas that inform a core philosophy of making.

Principles reflect underlying sensibilities for thinking through sound. They are markers of artistic creativity which embrace uncertainty, discovery, innovation, plurality, and an appreciation for communicative potential of sound via its many subtleties. *Principles* highlight sonic practices to be far more than basic craft actions, foregrounding the fundamental artistic processes at play when working with sound.

Principles are applied via Approaches and Techniques.

List of Principles

3.2.1 *Process of Discovery*

An unfolding procedural approach. Not just seeking to deliver an explicitly defined outcome, but allowing the materials and the process to direct the pathway to the final outcome, discovering it as it unfolds.

Embracing practice as a process helps us to appreciate the complex realities of making more deeply, reflect the journeys taken in realising a work more accurately, and celebrate the inherent creativity at play throughout the entire workflow.

Making is not just a simple realisation of a fixed idea, but a constant process of development, feedback, and investigation. Conventional ideas of making often simplistically separate out the 'idea' and the 'doing', compartmentalising the messy reality of practice into concept and action. Such divisions can also imply a separation between artistic inspiration and technical craft practice, which can, in turn, lead to the notion that some roles are more creative than others.

An unfolding *Process of Discovery* enables us to positively accept the integration of new ideas that emerge along the way (accepting that ideas and goals will evolve and change as the work develops), welcome new outside influences (embracing collaboration and dialogue as active contributors), and challenge outdated notions of creativity being separated from craft practice.

Steve Fanagan

"You don't start the job with a preconception that rules how you should work. You start the job with an open mind and you do all that you think you should, and then evaluate. You do that yourself and with the people around you. And hopefully, through an iterative process, you figure out a version which is truest to the essence of the story".

Peter Albrechtsen	"I often have a plan from the start, but maybe 20% of that plan actually works. Then I'll get another 20% when I see the dailies".
Hildegard Westerkamp	"[T]he life process leads me into a piece, and the materials into the meanings and structures of the work. This is very close to how life's journey is perceived in India [...] On a journey, it's not the destination that's important. It's the process of getting there that's important".
Trevor Wishart	"I set about collecting a lot of materials and begin to work with them to discover whether or not it's going to work! [Laughter] [...] [A]s you work with the materials, the form gradually begins to solidify around the materials. [...] I should add that I do all of this by listening. I don't have any formula such as 'golden sections' and all that nonsense. You do it by listening, and you understand [the] proportions because you experience them".
Natasha Barrett	"I don't know how much time I need to communicate the expression, and I might not even know what sound materials I am going to use. These things converge gradually over the months of working, throwing things out and finding things that seem to function. So, the plan starts to 'find itself' through the nature of the materials. The musical structuring decisions emerge during this process. You could also say that the overall plan emerges or converges".
Nikos Stavropoulos	"There is no overarching plan, but there are choices made at every step of the way".
Randy Thom	"[I]t's the process itself where the most interesting stuff happens. It's the discoveries that the artist makes, painting the picture, or doing the sound work, out in the field with a microphone, just listening. It's those discoveries that are, I think, most often what winds up driving the piece of art. [...] I think what we call creativity is more often actually discovery".

3.2.2 Constant Enquiry

A passionate drive to always seek out new combinations of sounds and new forms of sonic expression.

Innovation is driven by a passion for the new and a desire to constantly improve. Many of our conversations reflected such a drive, demonstrating that practitioners at the very pinnacle of their professions are constantly exploring new approaches to hearing everyday objects, finding new ways to convey emotions, feelings, and experiences via sound and still seeking to learn and discover new ways of doing things.

This quest for 'new' sonic possibilities at the highest level helps put to rest any assumption that there might be fixed 'golden rules' waiting to be discovered. If there were magic solutions, we would expect them to be held by those at the top of the field. Instead, we reveal that there are no secret formulas. Because there are many possible solutions to any brief, the only approach available is to engage in a constant search for the most resonant and effective solution for the task in hand.

Ann Kroeber	"Finding new ways to hear, [...] is what's really driven me. Because it's exciting. Because it's interesting. Always wondering about 'what if' and 'what next'".
KMRU	"My process is very intuitive and I'm always listening to discover what is happening when I bring different soundscapes together. I'm interested in creating new environments of sound".
Randy Thom	"[Y]ou have to keep your mind open to the unanticipated, the things that don't have any obvious connection to what the director has told you or what's in the script, but are things, in fact, that can be made to be very connected to both. [... T]hat's the creative mindset that I always try to be in. [...] Using something less obvious makes the viewer/listener go searching in their own minds to figure out what the meaning of that sound is, and how that sound connects to what they're seeing". [...] You have to have time to make some mistakes if you want to do anything very interesting, because if you don't have time to make any mistakes, then you're just going to duplicate something you've done before".
George Vlad	"[T]he more I travel, the more I explore and the more I listen, I realise that the world of natural sound is incredibly diverse. I keep discovering new things, new sound sources, new biophony and geophony".
Trevor Wishart	"Music is the impact of [a] formal structure, and we still don't fully understand how that works. [...] you're not *just* following these rules. A successful fugue has logic combined with its detailed emotional effect from moment to moment. That is what makes it something powerful musically which we can't explain, and you can't teach it to anyone".
Onnalee Blank	"[A] lot of Sound Designers and supervisors get stuck on their own ideas, 'this is how it's been in the edit the whole time and we're not going to change it' and I really hate it when people say that because, well, 'Why? Why won't we change it? Why don't we try something new? We're here at the mix to try stuff'".
Paula Fairfield	"I've always had a studio, I've always had my own gear, I've invested in myself over these years [... and this] allowed me to be in this constant state of exploration, out of the sight of judgement and anyone. And that's given me, in a place where women had almost no control, a sense of control over my own journey. So, it's partly about pure survival within the time that I came up in the industry, but it's also about me being constantly curious as an artist".

3.2.3 Unified Soundtrack

Non-adherence to traditional divisions of music/sound/voice. To fluidly embrace all sounds as material equally.

Slicing up the soundtrack can make it easier to digest (just like a cake!). But there is nothing absolute about where those cuts happen, and once made it can be difficult to put the slices back together and appreciate the seamless whole.

Our conversations revealed that practitioners tend to think in holistic ways about the entire soundtrack, without making explicit divisions of sound into fixed categories such as types of sound sources, production job roles, or their relationship to layers of narrative within the work.[4] This diversity of thought is also reflected by the multiplicity in job roles, many of our Film Sound contributors work across editing, design, mixing, and supervision, while the practice of Electroacoustic Music engages with the entire sound process working directly with materials from recording to performance.

Randy Thom	"The aspects of music that we use to tell stories are harmony and rhythm and dynamics. And you get all of those with sound effects as well. [...] I think a lot of the aesthetic borders that are assumed, or set up, between music and sound design are artificial and based on dumb assumptions. And most of the composers that I know don't recognise those borders or don't salute those rules about what is music and what is sound design. And the same for my compatriot Sound Designers".
Onnalee Blank	"Nicholas Britell is a composer I've worked with for quite a few years, as part of Barry Jenkins'[4] projects. That's a really wonderful collaboration. We have a weekly conversation about the work we are doing for certain scenes, we share stems of our work in progress and we really try to write around each other. Sometimes, I'll say to him, 'I need a trumpet in this key that goes with the drums I've created for the sound design in this moment'. So, it becomes a mind-meld of ideas".
Nina Hartstone	"If I'm working on the dialogue, I'm only one piece of the puzzle, one component of the frequencies that are going to be heard in the mix. If you play one sound on its own, it will sound very different than if you layer it up with other sounds. [...] It's also important to think about the story and overarching narratives in the film".
Nina Hartstone (in conversation with Trevor Wishart)	"I used to be very, very focused solely on dialogue, whereas I'm now very much thinking about the bigger picture. It's not even just the sound, it's the sound and the picture as well – it all needs to work together in the way that it's finally going to be heard. We have to take it in turns – effects, dialogue and music. It's no good if we all do a gasp, then there's a musical sting, and then there's like some big effects. Somebody has to take this moment [laughter]. Or we have to stagger, it just doesn't work to have everything all at once".
Steve Fanagan (in conversation with Annie Mahtani)	"Often in offline [editing], music is used to support something that sound hasn't yet had an opportunity to help with. And if sound does the right job we might make enough space, or create a mood, that doesn't require something melodic [...] For example: Stephen Rennicks who is Lenny's long-term composer, is often the first person at a mix to say, "Maybe we don't need music here". And that can be quite rare. It takes a certain confidence to not feel like someone is rejecting your work by choosing to take it out. And being part of a team like that is really exciting for me".

3.2.4 Assemblage

The merging of body, material, technology, and process.

An assemblage is a fluid network of different types of components which create complex and non-hierarchical interrelationships; each contributing equally to the process in manifold ways. It is not built of fixed elements but is dynamic and constantly changing.

Our conversations repeatedly returned to working practices which reflected a dynamic coming together (and apart) of many different component elements – people, tools, ideas, perspectives – which contribute in complex and networked ways to the delivery of outcomes.

It can be easy to consider individual creatives as islands of ideas, separated from their tools and materials, as well as other creative collaborators. But this simple and individualistic picture of practice doesn't fully reflect the rich synthesis of making that draws upon the unique intersection of all of the elements which contribute to it.

Assemblage reflects the complex reality of production which engages both human and non-human elements in a merging of tools, processes, sound materials, practices, collaborators, and ideas.

Jesse Dodd	"It's a really beautiful meshing of two worlds – the creative side and the technical side. It's all about teamwork, and when I say team, I also mean the technique – the microphones, Pro Tools – everything that's involved must be in synergy to make it happen. [… W]e are all connected, people and tools. […] Others have told me about this when they walk into my bay, and this often tends to be because we're all working collectively. […] We're all a team. If I don't do my job well, this affects the rest of the sound team, you're only as strong as your weakest link. It's a group effort, and I have so much respect for everyone".
Paula Fairfield	"I had this realisation when I was working on one of the *Lord of Rings* creatures. I was in such a 'zone' and I had this weird moment when I pulled outside of myself and I saw myself rearranging things, pulling stuff in and assembling things as though I knew exactly what to do. And I thought, 'How profound. How did I know how to do that?' Then I realised that I understood these feelings because I had been around people who had hurt me. I have that sensitivity, and this has been my way of processing it".
Onnalee Blank (in conversation with Nikos Stavropoulos)	"With *Underground Railroad*, Barry [Jenkins] holds on shots, directs and cuts his story for sound and music, which really allows us as Sound Designers to take the opportunity to do something cool. And it's always changing".
Steve Fanagan	"[I]t's not just my work; it's the work of a sound team, in collaboration with our Director and Film Editor. So, you're getting lots of clues and ideas from other people. [… But] there are so many people who work on the same films as I do whose path I never cross, because [in sound] we're coming in at the end of the process. Yet, somehow, we

do collaborate with each other. We have an intimate relationship that perhaps neither is consciously aware of, but you do still inform each other's work. Tiny details of the art department and production design, entirely influence choices I'm making".

Annette Vande Gorne

"It's important for me to have a gestural approach to control. [...] which allow[s] me to just perform and play with the space. And that is really important to me, because I was primarily a pianist in my training, and so I have this strong sense that music comes from the body. And I really need to have gesture. The interfaces that we use to control computers, keyboard and mouse, are really terrible interfaces for gesture".

3.2.5 Plurality of Potential (Multiplicity)

One sound may have many simultaneous meanings.

It can be easy to think of sounds as simple and direct markers, especially if we define sounds purely in relation to their source (e.g. "that is the sound of a door"). But sounds are not mere markers to objects which have fixed meaning. All sounds can shift their communicative potential as their context is modified (both in terms of combination with other sounds and images). All sounds convey deeper meanings via their characteristics (e.g. trajectory, timbre, balance etc.) and their position within the structure, contributing to environments, narratives, and building multi-layered meanings.

Ann Kroeber	"It's amazing how you can have the same original sound and use it for very different things. And you wouldn't even believe that it's the same sound!".
Natasha Barrett	"Sound has many facets and meanings, especially when separated from a visual source. It is an acousmatic result that has meaning beyond the spectrum".
Paula Fairfield	"A little sound inserted at just the right moment changes the story, or creates sympathy, or compassion where maybe it was never intended before".
Randy Thom	"[I]f you take, let's say, a fairly well-known sound effect that has been used in quite a few films, but you use it in an entirely different way, associated with completely different visual images than it has been in the past, then it's for all practical purposes, a new sound because it'll be perceived as a new sound".
Annie Mahtani (in conversation with Steve Fanagan)	"Sounds have to be saying something to me. I listen out for a combination of natural sound qualities (already present in the materials) and future potential (what can I manipulate and draw out)".

3.2.6 Clarity of Meaning

Nothing superfluous, only what you need. A form of essentialising which employs only the most salient auditory cues, in which you get to the most core and essential requirements without the unnecessary.

As Robert Bresson states in his *Notes on the Cinematograph* it is the filmmaker's job "to know thoroughly what business that sound (or that image) has there" and to ensure that there is nothing superfluous, to trim off 'all of the fat' (Bresson 1986, p.36).

This idea is also explored by Tim Nielsen in a blog post *On The Art of Economy*, within which he quotes the author Antoine de Saint-Exupéry's maxim, "Perfection is achieved, not when there is nothing more to add, but when there is nothing left to take away" (Saint-Exupéry in Nielsen 2011).

Randy Thom	"People often comment about my work that there is this focus, where you are very clearly hearing one or two things at a time. And the trick, of course is to do that while maintaining the illusion that the audience is hearing everything they're supposed to be hearing. I always try to arrive at as much sonic focus as possible. And I try my best to avoid cacophony unless the moment really calls for it, though that's not necessarily to say I avoid sounds that are cluttered".
Annette Vande Gorne	"I might choose to try and convey a particular sensation of a season, and in doing this, the goal is not to choose everything, but a particular moment, which the listener can understand immediately. And I don't mean this in the intellectual sense, but the feeling".
Peter Albrechtsen	"The more precise and impactful the sound is, the more emotional the sound is, the more it connects the audience and the more the audience connects to the film. […] It's very important for me that the film makes you listen and not just hear. There's so much noise in our everyday world that we constantly filter out, just to be able to survive psychologically. So it's important that every sound in a film is controlled and carefully designed to ensure that there are no superfluous elements which might trigger someone's brain to 'filter out the noise'".
John Young	"[I]t's like poetry, which] is typically very condensed and essential, it's about essences, which I think should leave you with a wider resonance. And I think that's the sort of thing we can do with electroacoustic sounds".
Natasha Barrett	"Sometimes I think that I have made an idea incredibly clear, but when I come back to it after three or four weeks, I realise that it is barely audible [laughter]. So, there is a long, iterative process of balancing clarity without making something blatantly obvious".

3.2.7 *Emotion/Feeling/Affect*

Eliciting an emotional response to connect with the audience via listening.

Communication of emotion and feeling sits at the heart of creative expression with sound. Across a wide range of our conversations the affordance of feeling was cited frequently as a driver behind a wide range of creative approaches, from microphone placement in recording, through to editing and mixing.[5]

Ann Kroeber	"It all comes back to this approach of feeling where to place the microphone and finding new perspectives on sound, placing the microphone in a way that allows you to hear differently in a way that you haven't heard before. [...] Find the place that moves you the most".
Annette Vande Gorne	"I might choose to try and convey a particular sensation of a season, and in doing this, the goal is not to choose everything, but a particular moment, which the listener can understand immediately. And I don't mean this in the intellectual sense, but the feeling".
Peter Albrechtsen	"I think the meaning comes through the feeling [...] Every sound I edit into a film has to tell a story. Why else would it be there? If you have that approach to sound, then it also heightens the audience's emotional attachment to the story, because the audience feels that everything is telling them something important".
Onnalee Blank	"I'm always thinking about [emotion], 100% of the time. Maybe that's just me, but anytime I'm working on a scene I'm always thinking how does this make me feel? What am I supposed to feel? What am I supposed to let the audience feel?".
Trevor Wishart	"Music is the organisation of sound in time which has an emotional effect on people. [...] Whether the music is interesting [or not] is something to do with the way that music affects us".
Steve Fanagan (in conversation with Annie Mahtani)	"People don't think enough about the fact that microphone placement (and point of audition) is not just technical, it's always emotional".
Nina Hartstone (in conversation with Trevor Wishart)	"If it's supposed to be an emotional scene and something makes me feel, I will definitely use those performances rather than anything else, even if another may be technically slightly better".

3.2.8 Sonic Communication

An acute awareness of the power of sonic subtleties to convey impressions and meaning.
The ability to listen intently, discern, and recognise the subtle details of sounds which might seem obvious and widely appreciated (attested to by the plethora of 'ear training' programmes available), but recognising the subtle potentials for communication elevates this category. For instance, it is one thing to understand and identify the equalisation (EQ) balance of a sound, but another to recognise the communicative potential that this balance might have for listeners.

Connected to notions of Emotion/Feeling/Affect (*Principles*), sounds are considered more than just markers of their source, they are complex conveyors of meaning. Practitioners might not be consciously aware of the rationale for their choices, but they frequently respond to their feeling, training, and intuition to identify and use sounds which have the power to communicate to others. This may be in a direct visceral way, such as with an animal's scream, or it might be in the choice of a specific timbre of a door creak, which imparts its characteristic into the mood of the scene.

Hildegard Westerkamp	"There is a message in every sound that we hear. For instance, the sound of a car passing tells us a huge amount – possibly the mood of the driver, the surface of the road, quality of the tyres, acoustics in the street etc."
Ann Kroeber	"You're always responding to how that sound makes you feel. What does it do? How does it affect you? It's very important to be sensitive to that. [… T]he emotion that they [animals] can convey with intonation and expression is incredible".
Nina Hartstone	"I often listen to Automated dialogue replacement (ADR) takes and say to myself, 'they're not smiling enough'. It's not because there's any laughter on the voice, it's literally just the way that the face shape changes as the voice comes out. Somehow you can hear it in the voice".
John Young	"[w]hat I've tried to do in *Ricordiamo Forlì* is use musical sound as a way of aestheticising, trying to re-elaborate the experience of what is going on inside the person when they are speaking. The music is aiming to stand in for the memory that is underpinning the process of forming words, because when we speak, we are often thinking many different things that we are trying to channel into a statement. That's what I'm trying to do with sound. But not just sound, musical sound. Sound that isn't just itself. […] not just illustrative but relational".
Paula Fairfield	"[F]or a sequence in Lovecraft, the Shoggoths circle the main characters and they're belching really loudly, and they're all breath, they're loud and you can almost feel it. The viscerality of that yelling is, in a weird way, like a direct expression. It's like that earthquake becomes their own pain in a weird way. It's like what you're experiencing".
Peter Albrechtsen	"You have to be very personal in your communication about sound. If you really want to explore what sound can do, it quickly becomes a very intimate and emotional discussion. You really have to open up about how you're feeling and react to certain things, the visceral reaction of hearing a sound, and I find that really interesting".

3.3 Approaches

Aesthetic strategies deployed to articulate sound and evoke affect.

Approaches are strategies or methods for elaborating sound. They can be used to guide the shaping and articulation of sound for creative purposes. *Approaches* enact *Principles* and guide the application of *Techniques* to express ideas or meanings via sound. They are more tangible, feeding to a greater extent, upon the specifics of creative contexts and the sound materials in question.

List of Approaches

3.3.1 Affordances

The potential available within the materials and processes.

The opposite of Limitations *(Approaches)*, ideas develop through engagement with the materials. Affordance is the way in which characteristics of materials and tools frame the potential options available, and thereby inform or guide processes of making. The creative unfolding of a work is guided by the details and features of material, tools, and context within iterative processes of engagement.

Annette Vande Gorne	"[T]he characteristics of the sound inform the way in which I am able to process them. Their characteristics dictate their treatment. […] The reason for employing a transformation is not the transformation itself, so it always changes based on the context [laughter]".
Hildegard Westerkamp	"[T]he sounds themselves guide me into a piece and its structure. The sounds are the words and I'm using those sounds to speak about something. The interest lies precisely in that interaction between the recordings, the environment at that moment, and the work in the studio when you're working with those sounds".
Steve Fanagan	"I don't necessarily know at the start how I am going to balance all of the soundscape elements to create that mood for the scene. It's a process of discovery. […] You need to be open to the process and get to the point where the film itself is telling you what it needs. […] I find that openness to process and the journey of making really inspiring; it's exciting too! You don't start the job with a preconception that rules how you should work. You start the job with an open mind and you do all that you think you should, and then evaluate. You do that yourself and with the people around you. And hopefully, through an iterative process, you figure out a version which is truest to the essence of the story".
John Young	"[W]hen I get well into a piece, […] I feel the material is under my control with minimal effort … I know it's potentials and how I can shape it. That also often tells me when I have reached the end of working on a piece, because adjustments or new elements start to detract, and take the work into a direction that no longer 'fits'".

3.3.2 Limitations

To be inspired by the limits of the system and use these boundaries to inspire creative problem solving and creativity.

Limitations provide frameworks within which to work, helping to offset the rich multiplicity of potential afforded by the sound itself.

Because the full potential of sounds are practically unlimited, working within the confines of limitations can help to drive creativity and encourage critical self-reflection to reveal new creative solutions. This might be in the decision to use sound materials from one specific location, or to simply limit the level of processing and transformation to specific functions or effects.

Ann Kroeber	"Sometimes people get so into the tech they think that having all the latest gadgets is where it's at [laughter]. They think 'Let's add something else. Let's add something else'. And they never stop to think, actually maybe they don't need to add something more. Maybe it's about using the tools that they have already in a more sensitive way?".
Hildegard Westerkamp	"[D]uring the process of composing *Whisper Study,* using only three sources (which included a recording of an alphorn from *The World Soundscape Project* collection) and applying the same techniques, provided a framing and prevented me from getting too confused with the technological potential of the studio. Often this constant framing is required, and it's why I've never been drawn to abstract composition and electronic sounds, as there's a danger of getting lost in space, and then I would lose my compositional inspiration. There are certain techniques that I use quite consistently, as I am often overwhelmed by the multitude of processing tools available in the technological world. And so, I like to have boundaries in my approach".
Trevor Wishart *(in conversation with Nina Hartstone)*	"When I'm working, I'm very constrained by the formal notions that are guiding that particular project – for example, the notion might be, 'making a piece using syllables from an array of different languages' – but I am constantly trying to elaborate the emotional within these constraints".
Annie Mahtani *(in conversation with Steve Fanagan)*	"Yes, exactly as you talk of the picture rejecting the ideas, so will the characteristics of my sound materials limit and direct possibilities. You've got to understand the character of the sounds you have and know how best they will behave, either in articulation or through processing – how far you can abstract it".
George Vlad	"If I'm out recording with a surround rig I'll quite often also record in stereo, because I can make a decision to focus on specific sounds. I think making deliberate choices and limiting my options is very helpful".

3.3.3 No Fixed Roles

A fluidity in practice between recording, editing, and mixing sounds. An ability to appreciate the nuances and demands of all three, even if the practitioner is primarily located in one.

With the advances in technology there is increasing fluidity between post sound roles in film, particularly editing and mixing. Electroacoustic Music Composers often work as solo artists, valuing the ability to control and shape sound at every stage of the process. In both cases practitioners identified the value and importance of understanding the whole production process. This flexibility in creative thought, enables a powerfully fluid creative practice, reaching outside of narrow frameworks to embrace positively contiguous practices or activities.

Annette Vande Gorne	"Composition is not just about assembly but about expression and performance. So, the very act of capturing sounds is inherent to my process. I might have wonderfully recorded sounds given to me by excellent sound engineers, but they have no resonance with the project and its concept. If I make the sounds myself I am able to respond to the nuances and the character of the project idea and make creative decisions within the sound collection process".
Nina Hartstone	"If I'm working on the dialogue, I'm only one piece of the puzzle, one component of the frequencies that are going to be heard in the mix. If you play one sound on its own, it will sound very different than if you layer it up with other sounds. So, if you over process […] you can be doing too much treatment to sounds that can introduce artefacts and artificiality, and you risk impacting on the connection that this dialogue has with the picture".
Annie Mahtani	"When recording, I'm always thinking about composition as well. I'm thinking about sonic space – where I place my microphones, how close I want to be to tune into a particular sound versus how much of the wider space I want to capture".
Jesse Dodd	"[E]verything is as equally important […], there is nothing more important than the other […]".
Paula Fairfield	"If you're tired of designing, go mix. If you're tired of mixing, go edit dialogue. If you're tired of editing, go field recording, go on location".

3.3.4 *Embodied Practice*

Making is a fully embodied practice and, just conceptual. Embodied Practice foregrounds the body through thinking of sounds as more than just abstract inputs for the ear, but as a visceral experience via the whole body.

Sound can be felt by the whole body and not just the ear. We respond to the pacing and tempo of sounds in relation to our breathing, movement, and heartbeat – we are indissoluble from our bodies and think through our movements and sensations.[6] However, one of the most famous phrases in philosophy is Descartes, "I think, therefore I am" (Descartes 1637). The consequence of this assertion has been to highlight a division between mind and body, which echoes across time to affect how we imagine almost everything we do today. The mind/brain is envisaged as the seat of abstract knowledge – pure ideas – while the body is considered as a mechanical vessel for transporting our brains about. Computer interfaces can help enhance this impression of a chasm between body and idea, by distancing us from the materiality of sound, translating everything into abstract keyboard strokes and mouse clicks.

As a result, it is more and more important for us to embrace our practice as embodied, and to remember that sound making, and our experiences of sound, are something that draw upon our whole body.

Annette Vande Gorne	"I have this strong sense that music comes from the body. And I really need to have gesture. The interfaces that we use to control computers, keyboard, and mouse are really terrible interfaces for gesture".
Jesse Dodd	"Sound is a vibration. A type of vibration that you allow into your spirit, and your being affects your movement, consciousness – it affects everything".
John Young	"[I]t's the excitement of engaging directly with sound in the studio […] that somehow translates into processes of selection and a motivation to bring a work into realisation".
Onnalee Blank	"I always think, 'If I can dance to this and it flows, and it has grace and power', then it's working for the scene. Transitions from one scene to another should flow, whether it's a hard cut or soft, transitions are a big thing. It's the same in dance too. It's not necessarily about how high your leg is, or how much you're pointing your toe. It's really the transition from step-to-step that makes you a graceful dancer, and I apply that same principle with sound editing".
Ann Kroeber	"[I]t's more about being present […] to the possibilities and to your own feeling in the moment. […] don't get caught up in the 'rules'. Don't think, 'I'm supposed to hold it in this way, six feet away' etc. […] Put the microphones where they sound good. Just find a spot and just do it!".
Nina Hartstone	"Quite often the actors are very focused on the technical details in ADR. So, I have to remind them 'you look very relaxed and you're smiling in this one and I can hear that in your voice'. I'm not sure if that nuance is something that most people listen out for but that's something that I've built into my practice over years and years of working with this material".
KMRU	"I spent my time just being in this new sonic landscape, trying to understand what's happening in it by immersing myself in the location, which of course also includes the actions of all of the people who are inhabiting this place. The character of the space was more important to me than any individual sounds that I thought I was going to get from it".
George Vlad	"I think it takes growing up in a specific place to really understand it. […] [Y]ou have to be a toddler, you have to be very close to the ground, and you have to be able to touch everything, and to walk and fall and get to the ground, and to climb trees, and to do all these things that children do to experience it fully".

3.3.5 Authenticity

How a sound might feel right (authentic for its purpose), but need not be realistically correct.

Playing with impressions of reality to convey meaning, authenticity is about articulating subjectivity and feeling. A sound's relationship to its original source can be limiting to discussions of its creative potential. Framing the discussion in terms of authenticity (away from objective reality) offers more nuanced opportunities to discuss the storytelling power of sound.

Authenticity operates via a balance of context and communication and is a major component in creative action within both Film Sound and Soundscape Composition. Related to both notions of Reality (*Techniques*) and Emotion/Feeling/Affect *(Principles)*, sounds can warp and transform impressions of objective reality, using their apparent recognisability as a facade to mask their articulation via a range of different techniques to communicate meaning and emotion.

Randy Thom	"[M]y job is really mostly about feelings. It's not about real authenticity - who knows what real authenticity even is in a film?".
Peter Albrechtsen	"I prefer to talk in terms of emotional authenticity. Somebody has set the sounds to feel emotionally authentic, but not necessarily physically authentic".
Nina Hartstone	"[Y]ou want to make it feel as real as possible and you want to do everything you can to try and make it *feel* real".
Paula Fairfield	"If you feel something you're going to start to feel it's authenticity, you're going to start to open up to its possibilities as a storytelling vehicle".
Nikos Stavropoulos	"It's not about representation and it's not about identifying a source, it's about the materiality of sound, how that is perceived, and how that is felt. It's about the tangibility and presence of sound. That's what I mean by reality, rather than the representation".
Steve Fanagan	"You're always trying to figure out what the world sounds like in the film. It always has a subjectivity, it's not direct – 'see a thing, hear a thing' – it's rather – 'what would this character be hearing and how would they be hearing it at this moment in time?' – and that's very much dictated by story and drama, performance and editing".
Peter Albrechtsen	"What I'm building and shaping is definitely my interpretation of this world, but a lot of that work is just done intuitively. Afterwards, I may realise, 'okay, I did this because of that'. But primarily it is guided by feeling. It has to feel right, but it doesn't need to be objectively real".

3.3.6 Embracing the Unexpected

You cannot know what you don't know. You must be open to the unknown.

Positively embracing opportunities of the unexpected is a marker of flexible creativity over technical craft. Embracing error, serendipity and chance, the creative recognises these unforeseen variations as contributing to the richness of experience, accepting that uncertainty is a natural and positive feature of the creative process. The ability to integrate new and previously unknown creative inspirations within the creative process, demonstrates creativity as a Process of Discovery (*Principles*), over the technical realisation of a preformed concept.

Each performance of a gesture or scene has a random chance element of variation within it which might reveal new possibilities.

Randy Thom	"I often say: A great craftsman is somebody who knows how to avoid accidents. And a great artist is somebody who knows how to use accidents. And I think that really speaks to the most important part of creativity as far as I'm concerned. It's the accident".
Hildegard Westerkamp	"[Y]ou have to be prepared for surprises and the unexpected. [...] how much are we willing to open up to the environment, and how much is the environment giving us the opportunity to be open to it?".
KMRU	"When I moved to Berlin, it was quite a shock how different the soundscape was to that of Nairobi, it's much more silent here. I spent a lot of time just listening, to become more attuned to what's happening here, to welcome new sounds into my sonic vocabulary".
George Vlad	"Natural sounds can be surprising. You can often discover something completely new that you've never thought existed, for example, the sound of empress cicadas in Borneo at dusk. It's very complex, and the acoustics of the rainforest make it even more exciting".
Ann Kroeber	"[D]on't let your prior assumptions about what kind of sounds you need for the project limit you, be there in the moment. Be open to the unexpected sounds that you might discover. I think that's so important".

3.3.7 Sonic Storytelling

Using the subjective interpretation of sound as a tool to communicate and guide audience's listening.

Assembly, transformation, and transition of sound can be used to guide and draw the attention of the listener, applying sensitivities for Sonic Communication (*Principles*) to construct affective meaning and stories. Sonic narratives which are open to embracing the listeners' subjectivity have the potential to draw listeners in by connecting with diverse personal experiences.

Annette Vande Gorne	"What interests me is to guide the imagination of the listener. To lead their interior cinema, by triggering memories, inspiring their imagination, and creating a reaction in them".
Paula Fairfield	"I create these stories so that I have a little trajectory in my mind of a performance, reaction, or a series of events. [...] Within that sequence are a series of little hooks that you are using to hang your interpretation on. Humans are so geared to take a series of events and turn them into a story that we

can understand. [...] I'm not narrating the story or telling you what the story is, but I'm setting out a series of events and cues which articulate and give a natural cadence to that series of little moments and, if it's natural and feels organic enough, you will naturally attach *your* story to it".

Steve Fanagan	"We're not recreating an objective sonic reality, but fashioning a subjective soundscape, which is much more to do with how we want to convey an impression of the character's feelings".
Trevor Wishart (in Conversation with Nina Hartstone)	"Sometimes you listen and listen and think 'I know, that is right. I can't tell you why, but it is'. And I guess the proof of the pudding is that other people think it's right as well...".
Hildegard Westerkamp	"One thing in common is that all of my pieces somehow relate to what I'm doing, thinking, and/or has been happening – the lived experience. And so, this interest is strong enough that I want to speak about it and create works that help to communicate this".
Peter Albrechtsen	"I often go back and forth between using exterior sounds of a natural environment and an internal subjective soundscape from the characters' perspective. So I'm thinking, what are the sounds that are surrounding these characters, but also what are the sounds that would be inside of these characters? What are they hearing? What are they feeling? How do I communicate that with sound?".

3.3.8 Sound and Character

All sounds have their own character, which may be apparent to a greater or lesser extent. These characteristics can be used to construct relationships, transposed on to objects or people within films.

Artists will make choices on sonic materials based upon their potential to convey a sense of character. Emerging from a range of parameters, characterful sounds have a depth or richness which provides greater narrative potential. Markers may include instability or change within a sound source, or the character might emerge via the specific conditions of recording (see also Acoustic Space (*Techniques*)).

Steve Fanagan	"You're always looking for the personality of something. When I go to the library and listen for sounds, I'm always seeking character. For example, if you think about an air conditioning unit, it's not the steady rhythm that's interesting, it's the moments of disruption where a rhythm changes or there's a shift in the cycle. Those little bumps are the most interesting because they feel individual to that object. When you figure out where to place such objects in a scene, their dramatic character can make a big contribution to the story".
Annette Vande Gorne	"If I hear a sound – which is just a slow fade in and out – it says nothing to me, but if there is a more dramatic gesture

	– [ssshup-ding] – I know that this is triggering something. There is a certain vocabulary of gestures which can be understood".
Onnalee Blank	"[I] try to think of sound as a character in the film. Not just to put sound in the movie but to create a motif with sound design that works in the same key as the music score".
KMRU	"I think that the things around us have voices, or want to voice something, and so my role as an artist is to enable them to speak. How can I take this glass [picks up water glass from the table] and make a piece for it by allowing it to share all of the sounds it can make".
Randy Thom	"As a Sound Designer, I think one of the things you always need to get better at is not thinking of sounds, literally, because thinking of them literally is the easiest thing to do. And so, the better that you are able to train yourself at thinking in terms of metaphor, and of what kind of feeling a sound evokes, the better you will be".
Paula Fairfield	"Generally, when someone gives me a brief it says, 'This creature appears, runs around, eats some people, and then goes off'. But that's not enough for me, I want to know *why*. I have to create a whole story. It's a process of thinking 'how can we make that resonate a little bit more, to extend it beyond just a silly moment.' I suppose it's like creating a performance".

3.3.9 Openness

Appreciation for the subjectivity in listening, that each listener may have their own unique interpretation which must be respected.

Everyone listens differently, and this needs to be integrated into creative practices. There cannot be just one interpretation, but an acceptance of multiple possible ways of access. Connected to Sonic Communication (*Principles*), Listening Empathy (*Techniques*), and the Plurality of Potential (*Principles*).

Hildegard Westerkamp	"I want to share what I'm hearing. And what I'm experiencing is so exciting that I want others to hear it. There's again a sense of conversation".
George Vlad	"When I'm in a space, I usually have a local guide with me. I try to work with locals who have grown up there who can sort of present or show me how they feel about the place, and the most interesting aspects of the place, and usually it's a very interesting process from talking to people who have grown up in the one place and haven't really travelled much outside of it".
KMRU	"It's an individual experience for each listener and I'm not very focused on one specific intention that I want the listener to perceive from the piece of music. I'm always keen to leave it open and simply invite the listener in".
Onnalee Blank	"I think that collaboration is a really vital part of the process. I don't mix what I create. If I'm sound supervising and I've

made some of the sound design, I actually prefer to have an effects mixer come in and mix the things that I've prepared, because I like to get a degree of separation from the material that I've been working on".

Natasha Barrett   "I do not want to change the way I am working to accommodate listeners who might find my music challenging, yet I want to create an environment for listeners to find a place within".

Nina Hartstone   "I think that you get the best out of your team by trying to make sure that everybody feels heard. I want to hear everyone's opinion, it's nice to have that breadth of things, and you definitely get more creativity if your crew have a broader base of experience, than if you just have one type of person creating a film. You want your film to appeal to everybody in an ideal world. Your viewers are a broad range of people, so it's important to bring that alternative experience".

3.4 Techniques

Material practices through which ideas and approaches can be articulated in sound.

Techniques are the most concrete of the three categories. Yet they are not simple building blocks or instructional steps. They are material practices which can be articulated for specific e(a)ffect. Ways of approaching and working with materials in order to articulate creative intentions. These categories of process emerged repeatedly within our discussions.

List of Techniques

Choice of Material p. 47
Shape of a Sound p. 48
Phrasing p. 49
Impression of Time p. 49
Noise/Texture p. 50
Reality p. 51
Listening Empathy p. 52
Chance p. 53
Performance p. 54
Proximity p. 55
Macro-Micro p. 56
Acoustic Space p. 57
Transitions p. 58

3.4.1 Choice of Material

A foundation of creative potential.

The process of sound selection may take into consideration a wide range of features and contexts, seeking out specific sonic features responding to the materiality of sounds themselves – timbre, texture, gesture – and their potential to convey impressions to a listener within specific contexts and for specific purposes.

Annette Vande Gorne	"I feel that I am the first listener. I'm not even really a composer at this point, I am a listener who responds to what the sounds give to me. And from the very beginning it's all about the feeling and about what these sounds say to me. If they say nothing, I don't choose them! [laughter]".
Annie Mahtani	"The starting point is always the sound. And it's always sound that I've recorded. Through the recording process I'm listening out for characteristics within material".
Nikos Stavropoulos	"But it really has to do with what kind of sounds display a heightened impression of physicality. Of course, there's a lot of granulation and granular reconstruction going on. But I'm consistently drawn to sounds that have a high frequency content and transients. Sounds that are present and that appear intimate".
Trevor Wishart	"The technical reason for working with the voice is that it's a very rich sound source. It shifts its spectral characteristics from moment to moment, which makes it both technically difficult and technically interesting to work with as an artist, [...] There's also this slightly different reason. [...That] if you make a multiphonic with a voice everybody is interested. They all want to know, 'How is that happening? What does it mean? Is that person frightened? Why is that so scary? 'We have this very, very direct link with the voice".

3.4.2 Shape of a Sound

Morphology and Archetypal Forms.

All sounds have a shape defined by their profile (Attack, Decay, Sustain, Release) and timbral characteristics, but this scientific terminology can belie their phenomenological affect. Physical sound gestures emerge from the energy of action that instigated them, "the application of energy and its consequences" (Smalley 1986, p.82). As such they can convey forward these gestures, such as to convey an impression of the physical movement. A prime example might be the swooping glissando of a 'crashing aircraft' sound.

Annette Vande Gorne	"[T]he archetype is a good tool to communicate meaning directly. There are two main types – the first is entirely physical. For example, if you take a wave sound and loop it so that it flows back and forth, it conjures notions of rocking – perhaps as you might cradle and soothe a baby back and forth in your arms – this is an impression quite distinct from others such as bouncing or friction, scrubbing or shaking, each distinct physical archetypes of sound. [...] The second archetype relates to the sonic image".
Peter Albrechtsen	"If I'm making a horror movie, then I want the car pass to be a screeching car pass 'vr-iiiiiiinnnnggg'. And if it's a comedy, then it needs to be a weird car pass that has some funny story to tell, 'hhuh-he-he'".

Trevor Wishart	"When I came to *Encounters in the Republic of Heaven,* I decided to move onto the level of vocal phrases, looking for the musical content within them – for example, they have a pitch contour, they have a rhythm, some expressive or dynamic contour – and these became the musical motif, or a 'tune,' which you could play with and contort".

3.4.3 Phrasing

Collections of Sounds into Small Motifs and Aggregations.

Grouping sounds into small units of association which create relationships between their characteristics and can provide an emerging sense of structure for the listener. Establishing sonic correspondences such as leitmotifs,[7] phrases and motifs might be associated with musical materials, characters, or spaces, or unfold over a specific sequence or scene.

Randy Thom	"[I]n the first few minutes of a film, you're teaching the audience how to watch and listen to the film, setting up for them a kind of mindset that will be useful for them to figure out what happens for the rest of the film".
Annie Mahtani	"I'm always much more moved by something that I feel is taking me on a journey or is clearly saying something to me. A really important part of my composition process is in drawing out the musicality of sounds and getting them to speak in some way".
Nikos Stavropoulos	"The compositional reality of sound objects, and/or sound events, is that they are more often than not, very composite things".
Trevor Wishart	"[A]s you start to work with the material on a very small scale, you make little bits and pieces and you build them up and eventually you might make them into an event which is totally spectacular".
Nina Hartstone (in conversation with Trevor Wishart)	"You need to think space wise in the frequencies, but also the arc within the entire scene, you don't want to over egg the pudding. You always need to leave somewhere to go, and you want to find the ideal high point within a scene, because each will have its own arc".

3.4.4 Impression of Time

Articulation of the Subjective Passing of Time, Faster or Slower.

Sound has the power to shape our impression of time. Expectations can be established and subverted to give increased urgency, or relaxed calm.

Nina Hartstone (in conversation with Trevor Wishart)	"You build up the audiences' expectations about what is going to come next, but if you flip that expectation on its head, you can change the context of what happened before. You can be very effective with sound in these moments, it often actually puts the viewer back into an active listening mode and directs them to pay attention to the next part of the story".
Trevor Wishart	"One typical experience that I have when you come to the end is that events which previously seemed well formed, now appear

	too long. On its own the event was wonderful, but in the context of the piece it's just hogging too much time, and it's holding the piece up. So typically, you have to shorten individual events before the work comes together as a satisfactory whole".
Trevor Wishart (in conversation with Nina Hartstone)	"The way music *really* works is that you remember what happened before and you relate that to what is happening now. Musical structure is essentially playing with the memory of events – the similarities and the differences. There are only two things you can do in music. You can either repeat something or you can change it. All music can be analysed in those terms [laughter]".
KMRU (in conversation with George Vlad)	"[I] didn't want the recordings to conform to this linear idea of grid time, because what is happening outside is not fragmented into such a timeline, it's fluid".
Annie Mahtani (in conversation with Steve Fanagan)	"I have to detach myself from the recording experience and make the decision to frame specific elements from the recording. To give listeners a time and a space to be able to listen, you have to zoom them in really quickly, and frame the overall experience so as to provide a space for them to listen within".
Steve Fanagan	"[W]e designed it so that everything was a little bit calmer and more open in the mornings. I recorded in lots of ADR booths, mix stages, and studios to give a range of dry room tones and from these we chose a light, static, air, which we enhanced with EQ to boost the upper-mids and remove the low-end. [… A]s the day progresses, and Old Nick's arrival is imminent, the room is closing in on you. [… T]he panning gets narrower, the tonality becomes richer (more low-end), and we begin to roll off some of the high frequency and upper-mids, which makes it feel claustrophobic".

3.4.5 Noise/Texture

A Richness of Noise.

Noise is often considered as unwanted, a negative opposition to sound. Yet noisiness is a core feature of sonic timbre, and the characteristics of noise can add an impression of richness and texture to sounds. Since the digital revolution stabilised, analogue tools have become more desirable, prized for their random variations and the noise colouration which the digital sought to emanate. Many contributors described a conscious effort to utilise and apply noisy sounds in their work for creative effect.

Texture conveys tactile impressions, which convey notions of Reality (*Techniques*) and can invoke notions of Authenticity (*Approaches*).

Randy Thom	"[Alejandro] Iñárritu has this term he uses cockayanga, which is just a word that he made up. For him, cockayanga means a kind of dirty, noisy, maybe even distorted sound. […] He thinks that cockayanga lends a kind of life to a moment that you don't get from something that is pristinely recorded and pure and perfect. […] When you introduce a sound that has a high noise component, it helps the storytelling because the audience often feels that this sound must be 'real', because who would have chosen to put that sound in if they had a choice?".

Peter Albrechtsen	"I hate anonymous sounds, as I said before, I feel sounds need to tell a story. [...] Often, story comes from the texture."
Natasha Barrett	"Our emotions and experiences, all the things that we do, could be described as the dirtiness of life. Nothing ever follows a perfect description of how things should be. I feel like the dirtiness of life is very much a metaphor for the actual experience of life. Music is, for me, about counterpoint, about composing, and about putting things together. And when you put things together, you show their differences, or their contrasts, to avoid uniformity or a sense of grey. Recording a sound with clarity will capture its dirtiness."
Trevor Wishart (in conversation with Nina Hartstone)	"You record something in the street, and it has all of this dirty stuff in it and that's interesting, you can work with it. I'm not necessarily interested in trying to isolate something from its context. It may be that there is something very interesting about the 'other' things in the sound. Sometimes you want that extra grittiness of the material because it's part of what your piece is about".
George Vlad (in conversation with KMRU)	"It took me a long time before I could disengage it so as to appreciate the beauty or the ugliness of the sound textures happening. [...] I can listen to it and disregard the fact that it was a bit windy and the microphones were a bit noisy, I can just listen to it for the aesthetic value of that sound. And that was a big achievement for me because it helped me to transcend notions of clean/dirty, good/bad".
Nikos Stavropoulos	"[C]hoices often have to do with the idea of physicality, in gesture, but also in the type of materials and textures. I'm keen that the action feels believable, and as real as possible, because I find that makes the flow of the musical text more successful".

3.4.6 Reality

The Real vs the Abstract.

Transforming relative impressions of reality, shifting between greater naturalism to abstraction. Such transformations might be used to direct listeners to specific features in the soundscape, or used to direct attention to events or objects within film. For example, shifting from a 'real' to a 'dream' state via the application of a rich reverb, or the gradual introduction of a high pass EQ filter to draw attention to a 'high pressure steam pipe in the ceiling'.

Related to Authenticity (*Approaches*) the very power of this technique lies in our ability to shift and vary the listener's impressions of reality.

Randy Thom	"[O]ur ability to manipulate sounds is so profound these days that you can take even a fairly familiar sound and modulate it or manipulate it in ways that will retain the essential emotional value of the sound, but make it pretty unrecognisable for people who may have heard the original version of that sound before. So, I try to approach all sounds as raw material to be played with and manipulated and experimented with in all kinds of contexts".

Natasha Barrett	"[I]n composition I must do more than explore individual field recordings. I try to engage the material through different degrees of abstraction, which then implies a perspective to which the listener may find a connection beyond the face value of the sound".
Hildegard Westerkamp	"[P]rocessing of a sound to me is symbolic for what happens when we listen […] to balance what we hear in the world with how we interpret it".
KMRU	"When I'm making compositions, I'll generally use the recordings from one place, overlaying and merging them together to create new environments and hyperreal soundscapes".
Steve Fanagan	"I try to use my own experiences to guide my decision-making. […] You're doing these unreal things but it feels natural. Hopefully there is an honesty and truthfulness to that emulation, which recreates real human experience".
Nina Hartstone	"[W]e are trying to create a reality, and the sound is there to transport the audience into what you're seeing. Even if the characters are up in spaceships and surrounded by unfamiliar sounds, you still want the audience to feel anchored in reality".

3.4.7 Listening Empathy

A Desire to Communicate and Connect with Others.

An active consideration of 'other' listeners. This is about respecting other people's opinions and embracing them to powerfully inform creative decisions.

Tied to notions of Openness (*Approaches*), Sonic Storytelling (*Approaches*) and Sonic Communication (*Principles*), alternative perspectives can greatly inform creative practice, opening up new possibilities, and deeper opportunities for more effective communication.

Paula Fairfield	"We're unique and yet we share the same experiences with so many people so we all can relate in some ways, even though the stories are slightly different. […] I suppose it is all about helping us to understand the other. That's the thing it's about right? It's understanding the other, and how you never know what someone else is going through. The only way we get a sense is if there's a little glimmer of something here and there. You might get a hint of something that somebody is feeling in a certain moment, but the rest of it is hidden".
KMRU	"It's an individual experience for each listener and I'm not very focused on one specific intention that I want the listener to perceive from the piece of music. I'm always keen to leave it open and simply invite the listener in. […] I think people become more aware of the 'other' when we become good listeners".
Annette Vande Gorne	"In traditional composition, I only ever learned about the architecture of sounds and the work, never about the architecture of the listener's perception, and this is a big difference! Thinking about the listener changes the way you think about and approach composing".

John Young	"I've always liked Stravinsky's expression that he composed for himself and the hypothetical other. [...] I hear things in sound that I want to coordinate and understand in my way, and shape into a form of expression that others might relate to".
Peter Albrechtsen	"Because sound is subjective, if you have a room with ten people and you play back a sound, you're going to get ten different opinions of what that sound means. There's no objective truth about sound, it is totally subjective, which is quite amazing. That makes it hard to talk about sometimes. But it's also really wonderful because it means that you have to be very personal in your communication about sound".
Jesse Dodd	"Interpersonal relationship is everything. You have all of these pieces and they're all equally important".
Steve Fanagan	"If you choose to engage other people in your work, whatever they tell you about it has meaning, purpose and significance. You can decide 'I don't care, I love it'. Or you can think, 'Maybe I need to interrogate this more'. My feeling is that the latter is healthier for your own creativity".
Nina Hartstone	"Everyone experiences the world through their own eyes and ears and we can only have our own impression of things. This is why I really love having a team and having fresh ears, because it's so valuable to try and get more than just my own subjective opinion on what does and doesn't work. [...] Your viewers are a broad range of people, so it's important to bring that alternative experience [...] If everyone's got the same perspective, it becomes very, very boring. There are fewer creative discussions. So, I don't think it's best for your final output".

3.4.8 Chance

An active use of uncertainty.

Chance can be used to discover new possibilities which transcend expectation or convention. The embrace of random functions, variation or aleatoric processes, to deliver and reveal new and entirely unexpected associations.

Randy Thom	"[O]ne of the first things that I do when I start working on a film is just randomly listen to sounds in a library [...] and inevitably, within a few minutes, I find a sound that I never would have anticipated. It never would have occurred to me to put it on a list of sounds to look for any of the scenes in the film, but it's a sound that has really interesting connections [...], to something that's in the script or some bit of action. And I think it'll be powerfully metaphorical or enhance some more obvious sound that I might find later to use in that sequence. I think that a John Cage style, chance operation approach for film sound can be really, really useful".
John Young	"[C]hance is a useful way of learning about what you can listen to. [...] I often feel that there is a kind of sound that I'm always

	looking for. However, I don't necessarily quite know what it is or how to get it".
KMRU	"[T]he tools that I'm now using include random or spontaneous functions and if I make mistakes or abstract changes sometimes I decide that it is better to embrace and keep these 'accidents' in the work".
George Vlad (in conversation with KMRU)	"I captured a recording of rain in the Namib desert by putting microphones on the branches of a tree, and when the soft rain eventually came, the drops hitting the wood made it sound almost like an instrument being played. The wood was so resonant it sounded like a marimba. I could never plan for that, it never rains in the Namib. So it's just a question of doing it enough times and eventually chance brings something into my recordings. I love this element and I'm trying to disconnect from trying to control everything".

3.4.9 Performance

Capturing liveness through real-time recording of gestures.

The vitality of performance can make a powerful contribution to the character and nature of sounds. Sounds are the result of physical action which has motion and gesture. Thus, working performatively (or thinking about the process of recording as performance) can deliver a vitality to sounds across capture, editing, and mixing. This is a practical realisation of an embodied way of thinking about sound.

Annette Vande Gorne	"Composition is not just about assembly but about expression and performance. So, the very act of capturing sounds is inherent to my process. [...] It's a real shame that the standard computer interfaces of the mouse and the keyboard are so limiting. I don't think it's by chance that engineers have conceived and designed software to reflect the analogue, for example with faders that give you the idea of touching something physical. But sometimes that ideal of gestural, tactile, control is missing".
Onnalee Blank	"When you're mixing, you're always performing, [...] I realised it was like being on stage in a weird way".
KMRU	"[W]hen you go back to record, you can almost feel like you're a part of the environment. You understand the choreography and the flow of the soundscapes, and you can move fluidly around to collect them. It's a much better situation than going into a place and hitting record without really knowing what will happen".
Nina Hartstone	"[I]f somebody is experiencing heavy emotions, they're actually breathing less. [...] whether it's pain or upset or fear, everything in your body is trying to just hold, and so the sounds aren't loose, they're escaping. [...] It's tempting for people to be melodramatic when they perform that type of stuff, but if you see someone genuinely upset, you'll see that most of the time they're trying to manage the sensations inside their body at that moment".

Jesse Dodd	"[T]he whole-body positioning is important. [...] The voice becomes a musical instrument; part of the scene, separate from the actor".
Steve Fanagan	"There is no getting away from the fact that one of our main goals in Film Sound is to present the performance of the actor. When a director and a film editor make a choice on a shot or take, it's more about performance and feeling, not necessarily about audio fidelity".
George Vlad (in conversation with KMRU)	"On my trip to the Amazon in 2019, I took a canoe ride at dawn on a tributary in the rainforest and it was impossible to not be in the recording. There's the movement of the canoe through the water, the sound of the wooden oar hitting the wood of the canoe, the sound of the oar hitting the water and splashing, my own movements and the boat creaking. Listening back, it's probably the most alive of all of the recordings I made on that trip, because otherwise they're very clinical, as they're recordings of a space, and there's no human element. And so, there's something that I like about me being present to a small extent in the recording itself".
Nikos Stavropoulos	"[C]hoices often have to do with the idea of physicality, in gesture, but also in the type of materials and textures. I'm keen that the action feels believable, and as real as possible, because I find that makes the flow of the musical text more successful".
Natasha Barrett	"Being performative often means moving either the microphone or the source. [...] [I]n the studio I often move the source. Outdoors, that's more tricky! When I am recording environmental sounds, I would describe it as more exploratory. Often the microphone is stationary and the landscape is the performer".

3.4.10 Proximity

The relative distance between perceived objects.

The apparent proximity of a sound can be conveyed by volume, reverberation and EQ. Related to Acoustic Space (*Techniques*), creating associations and relationships with the listener via the articulation of perspective.

Randy Thom	"Personally, I think that the most drama is to be had in how close a sound is to the listener. The proximity of a sound relative to other sounds or the changing proximity over time can make a huge difference. If you hear a lion growling that sounds like it's 100 feet from you, that's a very different emotional experience than hearing a lion growling that's three feet from you [laughter]. So, I think that dimension has the most power in it, in terms of playing with sound, how close a sound is to you".
Annie Mahtani	"When recording, I'm always thinking about composition as well. I'm thinking about sonic space – where I place my microphones, how close I want to be to tune into a particular sound versus how much of the wider space I want to capture. [...] We

think about sounds that are foregrounded and those in the background. Sound trajectories, morphologies, and gestures – sounds that emerge, sustain, and disappear. How sounds move and shift in space and time. All of which we can use as inspiration and as a guide in composition".

Steve Fanagan   "[W]hen Connell and Marianne first kiss, they get close to each other, and the sounds of the world drop away. For the kiss, it is almost silent. As they got closer and closer to each other, we slowly shifted the mix – remove the birdsong, take out the ticking clock, add more breath and cloth movement – and it begins to convey an impression of proximity and the feeling that you're totally lost in this moment with these two people. […] The world around has just got quite small and intimate; it's just about us".

Ann Kroeber (in conversation with Hildegard Westerkamp)   "We wanted to capture the sounds of galloping from the horse's belly, getting that perspective very close to the horse's hooves. And that really has worked well. So, it's different techniques for different purposes".

Nina Hartstone   "I would love to hear them breathing, but it's less important to me than the feel we want in the wider sequence. If we hear them breathing here it's going make it feel smaller and focus us down onto tiny details, but we're in a large-scale event, so having silence here contributed to that sense of scale".

Nikos Stavropoulos   "Smalley uses Edward Hall's proxemic classification of distances – social space, private space, personal space, and intimate space,[4] very effectively to discuss how we interact with sound in this context. Music rarely comes into what we perceive as intimate space, but it is very exciting when it does. […] Proximity is what happens physically, and intimacy is what happens psychologically. And I do prefer sounds that display those characteristics. I use them a lot".

3.4.11 Macro – Micro

Shifts of scale and perspective, large to small.

Using a relative sense of scale to construct sonic perspectives. Constructed within either the editing process from individual sounds, or guiding the capture of sounds during recording. Related to Acoustic Space (*Techniques*) and Proximity (*Techniques*).

Annie Mahtani   "I'm always thinking about perspective when I'm recording. In any session, I will usually make multiple simultaneous recordings of a sound or a space, which allows me to explore contrasting perspectives – from contact microphones, to ambisonic microphones".

Natasha Barrett   "An important difference between performativity indoors and outdoors is the time span. Outdoors involves greater time spans: the need to be patient and to let the landscape unfold at its own pace".

Steve Fanagan *(in conversation* *with Annie Mahtani)*	"[W]e create little mirrors or little breadcrumb trails leading into that [key] moment. For example, perhaps he steps on a floorboard or opens the door and we use a quiet version of that large gesture as part of these smaller sounds. The idea is to communicate that this sound gesture is so ingrained, and it's followed him through his life. It's a moment of schism, the shame and humiliation of getting caught and how his mother reacted to it, he's never really broken away from that. So, the moment has to be larger than life, and there's great licence in that".
Trevor Wishart	"Over the last–10–15 years my compositions have taken me from working with tiny fragments of voice-generated sounds, in *Tongues of Fire* – where they could be [makes a noise] noise, you know, any kind of extended vocal technique – to working with syllables in *Globalalia*, and finally to whole phrases with the piece *Encounters in the Republic of Heaven*. […] If you're recording isolated syllables, that doesn't matter, you can just treat the objects as sound objects. But with the complete phrases, I felt a responsibility not to mess unduly with what the speakers had said".

3.4.12 Acoustic Space

Sonic contexts within which other sounds are situated.

Acoustic Space refers to a context (ground), against which other individual sound objects (figures) might be situated.[8] It is therefore close to both Proximity (*Techniques*) and Macro-Micro (*Techniques*), but while these two pertain to individual sound objects, Acoustic Space is about the environment in which individual sounds are situated. Making use of acoustic reflections and delays, space might be built up to create an impression of scale, or collapsed in to create an impression of claustrophobia. Acoustic Space might be deliberately left sparse or empty to create a sense of unease or uncertainty.

Impressions of space, subtleties of acoustic reflection and delay details, might be conveyed by any form of sound system (mono to multichannel), enabling a perspective between sound objects or characters to be established.

Onnalee Blank	"[A] focal point of my practice is how to create a 3D sound with different mono files and panning and delays that spread your depth of field deeper […]. My feeling is that the field should be big and wide, and then you can really articulate the perspective that you want for the audience and the character".
Steve Fanagan	"Acoustic characteristics have a very clear emotional impact on me. We feel sounds; we don't think about them, we just instantaneously respond to acoustic properties, and they convey so much information about where the sound has come from – left or right, far away or close".
Nina Hartstone	"You're trying to get your viewer to empathise with your protagonist, so you want the viewer to feel like they are that protagonist, and that they are in that space, that they can hear the space around them".

Annie Mahtani	"Listening to find the right place to record from is the most important aspect. It's like taking a photograph, you've got to frame it so as to get everything in perspective and in proportion, so as to get a really full and beautiful sonic image of that space".
Nikos Stavropoulos	"This idea that you have a sound, and then position it in space, is perhaps a very visual centric way of working. In terms of how acoustic space actually works, there is no space where there's no sound. There is no sound event that exists outside of a space. That's not how it works in reality".
Natasha Barrett	"I am drawing on, and exploring, my own spatial experiences of the real world. I want to highlight these spatial features for other people to experience. So, I am still thinking about traditional electroacoustic ideas of textures, gestures, and movements, but manifested as spatial phenomena, exploded as a spatial experience in the music. […] I have used a crow or a seagull call in recent works. These are loud sounds that simulate reflections from the landscape and inform the space, which in turn connects to a degree of reality".
George Vlad	"[T]he acoustics of spaces fascinates me. I like to look for features in the topography of a place that influence how sound is perceived and how sound bounces around and is reflected. So, I like to look for gentle depressions in the ground where you can sort of get this parabola effect and look for tree lines that will offer really drastic slapback delay. I like to record at the edges of things, or where environments overlap because you get a bit more variation, a bit more variety there".

3.4.13 Transitions

Shifting from one sound to the next.

Articulating change, development, and transformation over time. It can often be easy to focus upon sounds themselves and forget the movement shifting between one sound and the next.

Onnalee Blank	"If every sound movement flows from one into another, it makes it feel like it was shot that way and the audience doesn't hear the edit. As long as it has beauty and grace and transitions are flowing, then I'm happy".
Steve Fanagan	"The idea that sound is static in a scene just doesn't make sense to me. It *has* to evolve because the experiential nature of cinema is as something that does evolve".
Nina Hartstone (in conversation with Trevor Wishart)	"One of the things that I really loved in your work was how sounds would transition. At the beginning I'd be very familiar with what the sound may be and then it transitions into a different type of sound, and it's interesting that you can do that not just by manipulating the sound that you're hearing, but the things surrounding it".

Onnalee Blank (in conversation with Nikos Stavropoulos)	"I'm a big transition person. Every cut should have a meaning. And I almost want to get a tee-shirt that says protect the cut. Any big event is a way to get out of something else without the audience knowing. Let me put something in to catch their ear and draw attention to something else, and then they won't notice that I'm subtly transforming this other thing. Whatever you can do to make it work smoothly".

3.5 [Blank Entry]

Insert Your Own Principle, Approach, or Technique.
Description ~ 50–200 Words

Example Quotation 1

Example Quotation 2

Example Quotation 3 …

Notes

1 These findings present a snapshot of creative practices that are necessarily framed by the contexts of practice at this moment in time (2021–2022) and the people we interviewed. However, we deliberately sought to engage a diverse range of practitioners from across Film Sound and Electroacoustic Music to ensure that the project reflected a diversity of opinions and insights, and many of our contributors.
2 *Principles* in blue/dark grey, *Approaches* in orange/light grey and *Techniques* in yellow/white.
3 While the map cites the *Principles*, *Approaches* and *Techniques* that we identified, there is plenty of space to enable you to add more 'themes' to the map should you see fit. This is not a definitive list but a map for exploration and discovery.
4 Each of these approaches is frequently used to segment and break up the soundtrack into more digestible chunks for discussion and analysis. But, this very process of segmentation influences the way in which we are able to understand the soundtrack. See Knight-Hill 2020a for a detailed discussion of this topic.
5 We had previously argued for the communicative power of sound in this regard and therefore sought to explicitly investigate this aspect via specific interview questions (see also Knight-Hill 2019).
6 See also: Emmerson, S. (2001a). From Dance! To "Dance": Distance and Digits. *Computer Music Journal*, *25*(1), 13–20. http://www.jstor.org/stable/3681631.
7 A leitmotif is a series of notes, harmonic progressions or rhythmic patterns associated with a particular theme, setting or character.
8 The terms 'figure' and 'ground' are key terms from perceptual gestalt psychology. The ground is a background context, which the figure is an individual object situated to the foreground.

4 Interviews

In the interviews that follow, you will find co-edited conversations, developed in partnership with each contributor edited for clarity and factual accuracy. Specialist terms are explained within footnotes so as not to disrupt the flow of the original conversations.

Each interview has been segmented with subheadings to categorise key topics, and highlight significant thematic ideas.

DOI: 10.4324/9781003163077-4

4.1 Film Sound Interviewees

4.1.1 Ann Kroeber

Ann Kroeber (b.1946) is a Sound Designer, Effects Editor, and Recordist who has become one of the most admired sound effect recordists in Hollywood. Ann's husband was the late Alan Splet (1939–1994), with whom she collaborated, adventurously exploring new approaches to recording sounds for iconic films such as *The Black Stallion* (1979), *Dune* (1984), *Blue Velvet* (1986), *The Unbearable Lightness of Being* (1988), and the *Dead Poets Society* (1989). Ann's sensitivity towards capturing evocative sound has made her library an invaluable resource for a host of Sound Designers working on some of the most iconic films of the last 40 years. Ann's sounds have featured in films, television shows, and games – including six films that won an Academy Award in sound, with a further seven nominations. The Splet-Kroeber library is currently available via Pro Sound Effects.[1]

Speaker Key:

I   Interviewer
AK   Ann Kroeber

Listening in the Moment and Feeling

I   *When we spoke before, you told me the story about how you started out working at the UN and you were recording sounds for Chinese New Year, and you discovered that you had this uncanny ability to capture the fireworks and the street sounds and to bring back the soundscape. Is there anything that you look for when you are out recording? How do you know when you've got the right sound?*

AK   What I try to tell other people is, get out of your head and not overthink it too much. Just be there and listen. The most important thing when recording is just to listen and to hear.

Find the place that moves you the most. Maybe certain sounds will inspire you and you'll go, "oh, wow, I can use this for that moment of the project". But don't let your prior assumptions about what kind of sounds you need for the project limit you, be there in the moment. Be open to the unexpected sounds that you might discover. I think that's so important.

Also, don't get caught up in the 'rules'. Don't think, "I'm supposed to hold it in this way, six feet away" etc. Just don't. Put the microphones where they sound good. Just find a spot and just do it! [laughter].

I think that you really need to be in a state where you're absolutely appreciating sound. There are amazing things to be heard when you just listen.

I   *So in this sense, are you talking about the thinking rather than the feeling?*

AK   Yes. Oh, absolutely. I do mean getting out of the thinking. Yes, the feeling is very, very important. You're always responding to how that sound makes you feel. What does it do? How does it affect you? It's very important to be sensitive to that.

Discovering New Ways of Listening

I   *Having said that, it is interesting in your practice that you have used quite a lot of new tools and that you've always been interested in finding new ways of listening, it seems to me?*

AK   That's true, yes. Using my Nagra[2] was such a great way to listen because you recorded what you've heard. And unfortunately, with the digital, that's not necessarily the case. You hear something and then when you go back and listen you think, "that's not what it sounded like to me at the time". You try to find ways to make it as close as possible. But I could never have done those fireworks[3] with a digital recorder because when you hit a certain level, it just cuts off if you go over. It won't record. But with the analogue, there's this little space where it does record. So that's how I got many of my most interesting sound[s].

It all comes back to this approach of feeling where to place the microphone and finding new perspectives on sound, placing the microphone in a way that allows you to hear differently in a way that you haven't heard before.

I   *What has driven you to try and find those new ways of listening?*

AK   That's a good question [laughter]. Just curiosity, I guess. Finding new ways to hear, I suppose, is what's really driven me. Because it's exciting. Because it's interesting. Always wondering about 'what if' and 'what next', and I think that's what drives you to discover new sounds.

I   *Are you thinking much about the audience at all?*

AK   No, not really. It doesn't seem appropriate. I just try to focus on the sound and capturing it. And then later you can figure out what to do with it, and how it'll work for people, and how it fits in with the film [laughter].

It's amazing how you can have the same original sound and use it for very different things. And you wouldn't even believe that it's the same sound! But there you go.

Emotion, Intonation, and Expression of Animal Sounds

I   *On the topic of animals, you seem to have a special affinity for recording and capturing the articulations that they have.*

AK   Absolutely. Yes, I've found that the emotion that they can convey with intonation and expression is incredible. I think one of the things I loved recording was the animals. And a really important part of this was being able to get out of the way and just let the animals talk to me. And I would talk to them!

I recorded a gorgeous Bengal tiger one time, Ceasar was his name, and he was really grouchy and frustrated. He was in a beautiful sanctuary for big cats, but it was fenced in. He could run around, but he was alone, and he was really cranky. He came over to me and growled in a gruff way 'grrr', as if to say "What do you want? You don't know how miserable I am", and I just told him what I was doing. I said to him, "Caesar, could I perhaps record you?" and I explained to him about a microphone and a recorder and the sound. And I pointed to my ears, and the sound goes in there. And then he just talked to me, right into the microphone. It was just incredible. He just talked.

And then at one point, we were having this conversation back and forth. He'd go, "Grrr, oh, you don't know the trouble. It's just so awful here. You can't imagine. Grrr". And then at one point he put his head against the fence and he wanted me to pet him, and I wanted to pet him so very much. Oh my gosh I wanted to so much, but that was something I was told profoundly not to do. And I just told him, "Oh, baby, I'm so sorry. I'm not allowed to", and he pulled back and he just glared at me. He just looked into my eyes and then he shook his head as if to say, "You don't trust me, do you?". And then he just walked away in a bit of a huff, as if to say "If you don't trust me then I'm not talking to you anymore" [laughter].

But it was just an incredible experience of connection. After using a lot of Caesar's sounds, I thought about him a lot and I wanted so much to go back and to pet him! I felt like, I don't care if you bite my hand off! I just want to put my arms around you and hug you. You're just such an amazing animal! [laughter].

But it wasn't just Ceasar, I've had that kind of incredible experience with lots of animals. One time there was a dragonfly and I said, "Come here, come here. I want to record you", and the dragonfly came up to me and it flew right into the glass that I was holding up and it really made this most incredible buzzing sound. It was just so beautiful. I was like, "Oh, my gosh".

They're smart. They're really just so happy if you respect them. They're just delighted that you try.

I   *It's something about that communication through non-linguistic means. Beyond words, through those intonations, those gestures, and those emotive communications?*

AK They have a real sensitivity towards understanding through the emotions and the feelings, and they really do respond to that. Animals are really something.

Sonic Intimacy and Presence

I   *It's quite intimate then, the relationship that you have with sounds? You like being close to sounds?*

AK Yes. I think so. Or far away if I'm listening to something and I'm on a hill on a mountaintop, coming down... so maybe it's more about being present there, really present. Present to the possibilities and to your own feelings in the moment.

It can be a problem to think about it too much. You just... I don't know... If you overthink it, you miss a lot. I think sometimes people think about the sound, and they don't just feel it. And they miss out on that immersion in the sound world.

For example, one of the things I provided for Paula [Fairfield] was the sound of a metal oven rack being pulled out. I'd FRAP[4]-ed it and it just screeched incredibly. And then slowing it down it made such an incredible dragon noise. It was really, really good [laughter].

You never know what you're going to get [laughter]. And when you're recording it, you might not be thinking 'dragon'. But then you find out that that's what it is later on.

I   *Could you talk a little bit more about your use of the FRAP?*

AK The FRAP is a curious thing because you can never know really what you're going to get. It's impossible to know what's inside, and what it's going to make when you have that microphone on. It's really an adventure.

Mostly it's just a process of trying different things and seeing what happens. I think the first time I was just trying it out on different things in the office that my husband and I were working in. There was this ventilation duct overhead and you could just barely hear the sound of the air rushing out of it with your naked ears. And then when I put the FRAP on the metal edge of the ventilator and turned it on. Oh my god, it was incredible. It was like five floors of sound coming through the shaft. It was an amazing variety of things, and you'd never expect that from just listening with your ears.

And that's the exciting thing about using the FRAP. It's just this whole world of just trying things and not thinking about it, just seeing what you can get. Try and you'll learn. At some point you learn that "This type of object will make a certain sound", or "It's better to put it on metal". There are certain things you can't put it on because it doesn't vibrate inside. For example, if you put the FRAP on a stone nothing's going to happen there. But those are things you just learn.

The sad thing about my FRAP is that I used to have this special wax that helped to bind the contact to the object that you were recording. The amount of wax could change the whole sound. It was a special medical wax that the inventor of the microphone, Arnie Lazarus, had and when he passed, I couldn't get it anymore. So I had to stop using it because it wasn't the same anymore. It was just so sad. I've tried lots of different things like beeswax, but nothing's really worked like that original gel. Oh, gosh, that stuff was so great. It was the magic of the FRAP. But I couldn't ever find the exact kind.

Arnie actually custom made my FRAP for me, so it was pretty special.

I   *Did you go to him with a specification for it?*

AK   Yes, we worked together. I heard sounds that he'd recorded and I was very excited about them. We were working on the pre-production for *Dune*[5] at the time, and I thought it would be useful to use that microphone to record new things and get new worlds for it.

Dune has so many industrial and unusual sounds, and so I said, "well, we should try that new FRAP" and gosh it was amazing. Alan [Splet] was just blown away when I recorded that ventilator, it was a real "oh my god" moment. So, it became a regular tool for that film… Gosh, taking that thing into a factory was really exciting.

It's just capable of such amazing things. Because in a way it isolates sound too, so you just hear the inside of something. I think nowadays more people use contact microphones, but the quality of the FRAP was really special.

Technology Is Just a Tool. Listening Is Key

I   *Your curiosity about sound is really inspiring and infectious, always going out and hunting for new sounds and asking, "What if?", "What might we do differently?". So, it's experimental in a way.*

AK   It is, very much. But, in terms of technology you don't need to have all the bells and whistles, it's important you have something familiar that works and great ears.

Sometimes people get so into the tech they think that having all the latest gadgets is where it's at [laughter]. They think "let's add something else to my kit bag. Let's add something else". And they never stop to think, actually maybe they don't need to add something more. Maybe it's about using the tools that they have already in a more sensitive way?

Sometimes very simple things can be beautiful.

I   *I suppose, if you think in terms of musicians, a musician will learn to play one instrument, and they will practice for eight hours a day on that one instrument. Whether it is the flute or the violin that instrument doesn't change; they learn to control it and articulate it in new ways. And I think sometimes, it's too easy working electronically to just add an extra thing or swap a bit out and use a different one…*

AK   Right, yes! Maybe we should just practice on the microphones we have and learn how to really use them? [laughter]

4.1.2 *Paula Fairfield*

Paula Fairfield (b.1961) is a Sound Designer and Sound Artist. Best known for her work on the iconic TV shows *Game of Thrones* and *Lost*, Paula has garnered numerous awards and nominations for her work in both Canada and the United States, including 10 Emmy nominations and two wins for her work on *Game of Thrones* (2015 & 2019) and for *Lovecraft County* (2021).

From TV, commercials, film, VR, installations, and other sound-related discourses, her passion is high concept sound design; over the course of her career, Paula has worked with auteurs such as James Cameron, Darren Aronofsky, Robert Rodriguez, Brian De-Palma, Paul McGuigan, and The Soska Sisters. Born in Nova Scotia, her work as an exhibiting video artist resides in several collections worldwide, including the National Gallery of Canada. In May 2021, she was awarded an honorary Doctorate of Fine Arts from her alma mater, NSCAD University. Paula became a Visiting Professor at the University of Greenwich in 2019.

Speaker Key

I	Interviewer
PF	Paula Fairfield

A Journey of Discovery

I   *You always talk about how you're on a journey of discovering and seeking new techniques. And it's this openness to this process of discovery, of constantly learning more about sound and learning more about how you can express yourself through it, which I think, is really refreshing. Often, people can be quite scared of admitting to that uncertainty and that even the experts are still trying to figure things out!*

PF   Well, that's my whole career. I never went to school for this, I have no formal training at all. So, my entire career has been teaching myself and discovering. I also have ADHD and I get bored easily. But, as I've experimented over the years I've realised that I could live five lifetimes of sound and never get to the end of it. It's kind of endless.

We are at a time when there are new tools being made constantly that are really amazing – the expansion of new immersive technology and so many different ways of exhibiting and displaying, and manipulating and digging into the core of the sound, pulling it apart, and putting it back together. It's a really interesting time to be working in sound.

I started on 16 MAG,[6] which is as pure as you can get. We had eight tracks when I started, and now we have 2,048 tracks available in Pro Tools! I remember when we went to 64 I thought, "what the hell will I do with all of those tracks?" But, fast-forward, and my system constantly crashes because I'm overloading it with way too many tracks! I'm constantly playing and having fun and trying different plug-ins and trying different sounds and what will work. It's freaking endless. So why not?!

I'm someone who always wants to try and do my best and I realised at some point that, especially as a woman, I'm always going to have to be better, I'm always going to have to have more up my sleeve. If I want more challenging projects, I have to continue to challenge myself because you're not going to get that stuff if you're not on top of your game.

So, I've structured my life in a way that helps me achieve this, I've always had a studio, I've always had my own gear, I've invested in myself over these years, and partially this was so I actually *could* learn, because as I came up through the industry I wasn't invited into the studios or labs where the sharing and discussions of new ideas were happening. But this self-investment allowed me to be in this constant state of exploration, out of the sight of judgement and anyone. And that's given me, in a place where women had almost no control, a sense of control over my own journey.

So, it's partly about pure survival within the time that I came up in the industry, but it's also about me being constantly curious as an artist. And I've been working in a medium which is the most curious and mysterious of all. It truly is. It's wonderful.

The fact that we have so many different disciplines within one medium, pure music to pure sound design, and everything in the middle, it's a testament to that. It's just endless. If you're tired of designing, go mix. If you're tired of mixing, go edit dialogue. If you're tired of editing, go field recording, go on location. There's no way you could master everything in this medium, I honestly don't think it's humanly possible to explore it all.

I   *Do you find it creatively inspiring to be able to explore these different types of sound practice?*

PF   What's fun about the space that I work in is that I *am* using these different techniques. I'm creating and trying new things all of the time, and then, I get to fine-tune them and edit them up against picture and see if I can convince you that that's actually what happened. That process is totally addictive for me, and I absolutely can't get enough of it. There is something about the combination, making something appear to do something that clearly it can't, but it sort of looks like it does. That's what's cool. [laughter].

It takes you on this journey of experience. I do love working against really good picture for that because I can get in there and do some crazy stuff for really good moments or sequences. And hopefully that enhances and lifts the shot, and sometimes tells its own story in tandem, which adds more richness to what's going on.

It's a fun space to play in, for sure.

Building Worlds – Sonic Storytelling

I   *I would love to talk a bit more about this idea of the parallel narratives that you always create within your work, and about thinking beyond the scene or the frame.*

PF   Well it's world building in a way, right? And world building is not just background, it's about creating pockets of activities that appear to have their own little sense of logic going on in tandem with the main event. Activities that add another interesting kind of comment, or flare, or insight; or even rub counter, in contrast.

And that's where I create my own framework. I'm often telling a story that is nothing to do with the main narrative. It's the story that I tell myself so that I know how to assemble the sequence, no matter what the creature is.

Generally, when someone gives me a brief it says, "This creature appears, runs around, eats some people, and then goes off". But that's not enough for me, I want to know *why*. I have to create a whole story. It's a process of thinking "how can we make that resonate a little bit more, to extend it beyond just a silly moment". I suppose it's like creating a performance.

I	*Is it possible to unpack and explain a bit about how you actually create that resonance and that connection with the audience?*
PF	I create these stories so that I have a little trajectory in my mind of a performance, reaction, or a series of events. And I build these from the natural cadence of life.

What I mean by this is, imagine you're in a shopping mall watching a group of people interact – people watching is a great thing, right?! – and as you do this you are attaching *your* story to what *they're* doing, which I guarantee has absolutely nothing to do with what is really going on for them.

Within that sequence are a series of little hooks that you are using to hang your interpretation on. Humans are so geared to take a series of events and turn them into a story that we can understand.

So I make *my own* story from these hooks. I'm not narrating the story or telling you what the story is, but I'm setting out a series of events and cues which articulate and give a natural cadence to that series of little moments and, if it's natural and feels organic enough, you will naturally attach *your* story to it.

For instance, imagine a creature rampaging which seems invincible. But when somebody hits the creature I attach a 'wince' or 'wail' to it. Suddenly, this conveys an idea that this creature is not invincible and was perhaps already hurt. In that millisecond moment you're going to suddenly change your mind about that creature, you're attaching a lightning-speed story to that moment which adds to that scene. It's going to register in you somewhere and make you think, "that creature isn't just random". It creates these little pockets of interest.

I didn't know I was doing that until I started to reflect on the first two seasons of 'Thrones'[7] that I worked on. I can tell you exactly what was going on across seasons three and four and what story I told myself for every single sequence. It doesn't matter that you know *my* story, what matters is that there are enough evocative moments that make sense in a sequence, which you will assemble according to *your* own story.

Part of my fun is making the story up in the first place, which gives me direction and guidance on exactly what I need to do. Once I've got a story going for a creature, I can easily identify the best moments across a sequence which I can highlight, to evoke the emotion, and reveal another layer that you never anticipated. Then you start to feel.

A little sound inserted at just the right moment changes the story, or creates sympathy, or compassion where maybe it was never intended before.

I	*I'm interested then in two things: one is about how you're drawing inspiration from real life, and the other is about how much you yourself are projecting into other people's projects.*
PF	Well it has to come from somewhere, right? So what I realised, from playing in this fantasy realm is that the idea of "one toe in reality" is really important as the bridge into supernatural and, as long as it's firmly anchored there for the viewer, this toe always allows that entry point. For me, there are straight up fantasy shows that have no toe in reality that I don't even enjoy as a viewer, but I always said, if I could convince you that dragons are real, I can take you on a journey because you've now committed. And part of that is through evoking an emotional response.

If you feel something you're going to start to feel its authenticity, you're going to start to open up to its possibilities as a storytelling vehicle. It's going to be more meaningful because you're going to start to insert your own story into it based on what you're feeling.

That's the area I'm really intrigued by. When I realised how potent that was, in my own journey of grief and recovering from different life events, I saw the effects of being vulnerable in the work. Even though I'm working for someone else's vision, it brings power to that work.

For example, one of the early scenes that I worked on for 'Thrones' had a powerful impact. It was the Plaza scene,[8] when [Daenerys] hands over her dragon in exchange for an army of soldiers. I had people say to me that it evoked sobbing in them, and when I started thinking about that, I realised that that scene was everything for me. It had everything in it. It had that little bit of joy, it had fear and anxiety, "No, no, no mum. No, don't leave me here" crying and shrieking, and it had vengeance and anger at the end.

I was having an incredibly difficult time at that moment, with complete upheavals in my personal life, and it's almost as if I was articulating everything that I was feeling in that moment through this sequence. I didn't consciously know that I was doing it, but I was responding to certain points in the sequence really strongly. I just heard in my head – hit there, hit there, hit there – and both myself and everyone else filled in all of the other blanks, associating them with moments that they recognised from their own life and made them feel the same way, I guess....?

I'm still not sure I entirely know how it works, but I'm having fun trying to do it! [laughter].

Viscerality, Emotion, and Feeling the Other

I   *Often with your creatures, and those fantasy elements, they have this viscerality and this real, genuine, organic animal nature. And, beyond the fact that they feel organically real, I think that they genuinely communicate something, they are affective.*

PF   During the course of my life my deepest wins[9] have often come from people close to me, in an intimate space, raging in pain.

You can have a number of reactions to that, but part of the healing process is to come to empathy and compassion for the one that has inflicted pain. When you realise that people who rage, and inflict pain, are themselves in deep pain, it changes everything.

I had this realisation when I was working on one of the *Lord of the Rings*[10] creatures. I was in such a 'zone' and I had this weird moment when I pulled outside of myself and I saw myself rearranging things, pulling stuff in and assembling things as though I knew exactly what to do. And I thought, "How profound. How did I know how to do that?" Then I realised that I understood these feelings because I had been around people who had hurt me. I have that sensitivity, and this has been my way of processing it.

I never understood, until that moment, that in every successful creature that I have created there's always a vulnerable little crack in them. They're scary monsters but they have a weakness. You see for a moment, here and there, their weakness or their own pain, and that brings out an empathy for these creatures that you should otherwise be afraid of.

And that's important to me because so many of us have been hurt by people who have inflicted pain on us. We feel our own pain, but when we can understand the pain of others, especially those who are inflicting it, it just brings a different sense of humanity and vulnerability.

And viscerality is part of that, because what I remember is not the sound, it's the feeling of it. I had an encounter once with someone who got so close to my face and was screaming so loud that it was blowing my hair back. I just stood there because it wasn't a new experience for me, but the feeling of that moment is etched in my soul in a weird way. So for a sequence in *Lovecraft*,[11] the Shoggoths circle the main characters and they're belching really loudly, and they're all breath, they're loud and you can almost feel it. The viscerality of that yelling is, in a weird way, like a direct expression. It's like that earthquake becomes their own pain in a weird way. It's like what you're experiencing.

I suppose it is all about helping us to understand the other. That's the thing it's about right? It's understanding the other, and how you never know what someone else is going through. The only way we get a sense is if there's a little glimmer of something here and there. You might get a hint of something that somebody is feeling in a certain moment, but the rest of it is hidden.

Sound Materials – Expression and Communication

I   *What's fascinating about this is the emotional focus and that emotional sensitivity, and that ability to express these kinds of experiences through sounds. To what extent is it important that you're communicating this emotion through sound?*

PF   For me it couldn't be any other way but sound. We're unique and yet we share the same experiences with so many people; so we all can relate in some ways, even though the stories are slightly different. For me, sound is the perfect medium for that. When you include the viscerality of it, *oh my god*, there's so much. It's a medium unlike any other, powerful in its ability to literally touch you with air. There's no other medium like it.

   Really it's a bit of a weird thing that I've ended up in this creature zone, but I see how powerful it is for me. Because these experiences I've had have affected me in a certain way, I've internalised them and now I'm able to respond to these as a human, and to draw on this as an artist and share them is a really interesting thing.

I   *Can you talk a bit about the materials that you choose to work with in order to articulate these emotional narratives, what draws you to particular sound materials?*

PF   I often use animal sounds, which for me is the right answer because I'm only interested in the pure expression of emotion. As far as I'm concerned that only comes from animals and babies. If you have an actor come in to give you vocal samples, they're acting that emotion, that's not real. I want the real stuff. So, I love to use animal sounds because they are truly primal.

Making Fantasy a Documentary

I   *I wonder whether it also links back to this comment that you had earlier about 'toes in reality'. I wonder whether it is about this world building? About putting little anchor points that can be drawn on?*

PF   It's almost as if you have fantasy but, in a weird way, you are attempting to make it appear as though it's a documentary. The comic book stuff is not that interesting to me personally because it's pure unadulterated fantasy. And that's fine for people who enjoy that. But I want to understand a little bit about *my world* from *that world*, and I like it when it could cross back over. So, it straddles this space, and it creates something that is challenging and interesting and impactful, because you're listening in a more raw and emotional way if you're engaged on that level.

I   *I love that idea of making fantasy documentary. And does that translate across to when you're out recording sounds, when you're out capturing sounds? I know you've done a lot of animal recordings as part of your work.*

PF   Yes, to a certain extent. I also love animals, and I'm really horrified about what we're doing to this planet and the beautiful creatures. The fact that we're destroying everything around us and this fact that humans are just awful as a species is a side comment that is constantly running through my work [laughter].

If I make a creature that is supposed to be scary and 'other', vulnerable, you feel the vulnerability and it transforms the impression from, "There's that creature, I'm going to kill it!" into "Oh but, that creature is hurt?!".

It's meant to evoke a sense of empathy. And I'm constantly in total childlike awe, over the sounds that creatures make. They're just so beautiful, all of them. One of the things I do is just listen in and listen out to the sounds of animals all of the time, looking for that little thing that I'm seeking. I'm listening to them talk to me all day long, and it's beautiful.

Field recording is special because I feel like I'm in this beautiful space, this beautiful home that mother earth, the ultimate artist has created, and I'm here with all of these creatures. When I get to record actual creatures, I'm always just deeply humbled by it. I also take those recordings with the utmost respect and privilege and honour, and try to make them meaningful in their roots.

For example, the death of [the dragon] Rhaegal in *Game of Thrones*, the shriek sound he makes when he is shot out of the sky is a Mississippi sound crane which has this beautiful call that many people haven't heard. There's 'something so poignant in it. I wanted to make this a moment where you felt something. Where you felt that this beautiful creature has been shot out of the sky, and there is something significant in its pain being expressed by another beautiful creature that is on the endangered species list because we've been shooting them out of the sky!

When I get the opportunity to do field recording, I feel it is such a powerful experience. If I had the opportunity I would just travel the world, give talks and record everywhere and that would be it. But I also respect so many independent recordists all over the world who are going places that I will never get to go because of time and work responsibilities. I love listening to the recordings that they make and how close-up they get. So, there's a lot of stuff to listen to and play with these days because we have a massive community doing work all over the world now and that never existed before. I just love that there is so much sharing within our community, for sure.

Final Remarks

I   *Thank you so much for sharing an insight into your creative artistic practice and for sharing how your human experiences inform the incredible work that you create.*

PF   I've been fortunate enough to have work that's allowed me to explore a lot of things about myself, allow to heal myself through my work, and make a living. And for that I feel so blessed and grateful, I can't even tell you. Because I don't know that I would be here without this work, and so, it's been incredibly powerful on a million different levels, and I'm excited. And that's why I want to lean into everything and always give it my all, because these are all opportunities on my journey, in my own life, to explore, expand, and grow.

4.1.3 Randy Thom

Randy Thom (b.1951) is a Supervising Sound Editor, Sound Designer, and Re-Recording Mixer based in the United States and is the Director of Sound Design for Skywalker Sound. His credits include *Apocalypse Now*, *The Incredibles*, *Return of the Jedi*, and *The Revenant*, working in creative sound capacities in over 175 films. He has worked with a diverse list of leading directors, including Francis Ford Coppola, Steven Spielberg, George Lucas, Robert Zemeckis, David Lynch, Guillermo del Toro, John Waters, Errol Morris, Henry Selick, Peter Jackson, Brad Bird, and Chris Wedge. With fifteen Oscar nominations, an Emmy, and a Grammy, he has received two Oscar awards: *The Right Stuff* and *The Incredibles*. He received the C.A.S. Career Achievement Award in 2010 and the MPSE Career Achievement Award in 2014.

Speaker key

I   Interviewer
RT   Randy Thom

Sonic Focus

I   *Do you feel like you have a particular aesthetic style that shines through different projects?*

RT   People often comment about my work that there is this focus, where you are very clearly hearing one or two things at a time. And the trick, of course, is to do that while maintaining the illusion that the audience is hearing everything they're supposed to be hearing.

I always try to arrive at as much sonic focus as possible. And I try my best to avoid cacophony unless the moment really calls for it, though that's not necessarily to say I avoid sounds that are cluttered.

For example, [Alejandro] Iñárritu[12] has this term he uses *cockayanga*, which is just a word that he made up. For him, *cockayanga* means a kind of dirty, noisy, maybe even distorted sound, and that is the kind of sound that he prefers. The type of sound he hates is something that has been pristinely recorded and seems pure, because for him that reads as artificial and not of the real world.

I think I had a similar way of working before I met him, but working with him made me realise it consciously. He really taught me to think about sound in a new way, and to get even further away from the idea that our job as sound people is to collect the best possible recordings of everything completely isolated. Because for Iñárritu, that takes the soul out of the experience. He thinks that *cockayanga* lends a kind of life to a moment that you don't get from something that is pristinely recorded and pure and perfect.

As a result, Iñárritu is often asking for more noise, but noise that serves the story in some way.

I   *We talked before about the idea that these cokayanga noises perhaps have some kind of history and story of their own?*

RT   There is a tradition that exists in American filmmaking of sticking as much as possible to pristine and clean sounds. So, when you introduce a sound that has a high noise component, it helps the storytelling because the audience often feels that this sound must be 'real', because who would have chosen to put that sound in if they had a choice?

So, I think that style does lend a kind of verité truth to a moment that it might not have otherwise. And that's one of the great things about Iñárritu's films.

Truth and Authenticity

I   *I'm really interested in this question of truth. At the heart of it is the fact that film is completely constructed and, I suppose, is always seeking to mask its own falsity in order to highlight the truth of the story.*

   Do you think that there's a certain characteristic or approach to working with sound that can help bring out this sense of authenticity?

RT   Of course, my job is really mostly about feelings. It's not about real authenticity – who knows what real authenticity even is in a film? – but oddly, I think using a sound that is oblique to the story, which isn't an obvious choice, can sometimes add more of a feeling of authenticity.

   Using something less obvious makes the viewer/listener go searching in their own minds to figure out what the meaning of that sound is, and how that sound connects to what they're seeing, and sounds that they've heard before, and other sounds that they may be hearing roughly simultaneously. I think that makes a moment or a sequence in a film feel authentic.

   I think the audience doesn't want to be spoon fed information; they want to be invited to participate in the storytelling. You don't want to confuse them too much, but you do want to seduce them, to entice them to participate in the storytelling. And so, a little mystery, a little bit of *cockayanga* sometimes helps in that regard.

Thinking via Metaphor and Feeling

I   *There is perhaps something in this to do with moving beyond literal simplicity and directness?*

RT   As a Sound Designer, I think one of the things you always need to get better at is not thinking of sounds, literally, because thinking of them literally is the easiest thing to do. And so, the better that you are able to train yourself at thinking in terms of metaphor, and of what kind of feeling a sound evokes, the better you will be.

   So, if you're working on a movie – for example, you're sound designing some race car engines – actually going out and making recordings of race car engines is fairly straightforward. It's time-consuming, but not very mentally taxing. What's much more difficult is to figure out how to stylise that sound, often by adding other sounds to it, like animal roars etc., and how to tailor those animal roars so that they have the same pitch dynamics of the revving engine, so that they tend to disappear into the revving engine, so that nobody in the audience is aware of this layering that you've done.

   But I'm always amazed at how much we can get away with, as sound people, in terms of what we present. The audience will tend to go out of its way to accept almost any sound we present to them and try to make sense of it in the visual context that they're given, and that's really one of the powers of sound, because you force people, in a way, to bring their own experience into the act of watching and listening to this thing, whatever it is.

Design Films with Sound in Mind

I   *In one of your blog posts, you mentioned how filmmakers could choose to shoot their film to better create opportunities for sound, and at the heart of that, I guess you're talking about the opportunity for that space, to make people wonder and to make people think.*

I'm really interested in this idea of drawing people into the film because I think your work is often very effective about drawing people in, and I suppose this links back to this idea of having just one or two sounds that people hear, but with a perception of a wider soundscape. It's then that sense of minimalism, if I can use that phrase, which draws people into that story.

RT   Yes, I think that's true. I think it's interesting that often the most intriguing or compelling sound design in films is at the beginning of the film, in the first few minutes. I think the reason for that is because that's the most opportune time to raise aesthetic and storytelling questions.

Often, in the beginnings of films, you're not exactly sure what you're looking at, you don't know who these characters are, and what this place is. And so it's at that stage of your film experience that you're all ears, so to speak.

You'll hear sounds but you're not quite sure what they are or where they're coming from. It's when the audience is most suggestible and so this often serves as a kind of playground for sound design to make a statement about the visual, or to make a statement contrary to the visual that will pay off later in the story.

In those first few minutes, you'll hear sound in a very subjective way, and as a Sound Designer, you're trying to seduce and intrigue the ear.

I   *Do you feel that these sounds are almost, in a way, posing a question? Sound has this possibility, this malleability to be more mysterious, and maybe then poses a question?*

RT   Yes. A very good example of it is the first few minutes of the Coppola film *The Conversation*,[13] which is a classic film for sound in many ways. But there's certainly some mystery about the visuals in that opening sequence when Harry Caul and his compatriots are trying to record a conversation in the town square. But the sounds are more mysterious, I would say. And so mysterious, in fact, that you often can't tell whether you're listening to sound effects or music in some of those sequences.

As one of the objects of the surreptitious surveillance walks in front of the camera and walks past the frame from left to right or right to left, you hear this sequence of tonal beeps and whooshes and noise that you eventually figure out is a kind of distorted version of what the person is saying. But in the beginning, you're not at all sure what it is that you're hearing.

And that mystery really serves the story, for lots of reasons, but largely because the whole film is about listening and not listening, and hearing and mishearing, hearing the wrong thing. And the mystery of speech and communication. And so presenting such a stylised version of the sound early on sets you up for that.

They say, in the first few minutes of a film, you're teaching the audience how to watch and listen to the film, setting up for them a kind of mindset that will be useful for them to figure out what happens for the rest of the film. And I think that film does that very well.

The Musicality of Sounds

I *Earlier we talked about sound having an emotional impact equivalent to music and I wanted to dig a little bit into this question and ask if you think there is a boundary that might exist between sounds and music? Do you think that there is a clear divide? Or is it something much more murky and mysterious?*

RT Oh, there are certainly clear divides in some ways. I often joke that it's not unusual in the final mix of a film for somebody – often a Producer or somebody who hasn't been involved day to day during Post-Production – to hear a particular sound that they're not happy about or confused by and they'll say, "what is that sound? Let's get rid of that sound effect". And very often, it turns out that it's part of the musical score! And suddenly they don't want to get rid of it anymore. They're converted to, "Oh, wow, that's really interesting" [laughter]. And that speaks to the hierarchy of traditional music relative to what we call sound effects or sound design.

So in terms of people's expectations, the lines between what we call music and what we call sound design are reasonably clear, but in terms of the actual aesthetic lines and storytelling lines, the borders between the two are much less distinct. The aspects of music that we use to tell stories are harmony and rhythm and dynamics. And you get all of those with sound effects as well.

For example, think of the incessant phone ringing at the beginning of *Once Upon a Time in America*.[14] Yes, that's absolutely a sound effect, but I think it has a fundamentally musical effect on people as well, not only because it's a tonal sound but because of the rhythm. It's like a very long drawn-out ostinato.[15] So I think a lot of the aesthetic borders that are assumed, or set up, between music and sound design are artificial and based on dumb assumptions.

And most of the Composers that I know don't recognise those borders or don't salute those rules about what is music and what is sound design. And the same for my compatriot Sound Designers. So, it's a struggle for all of us to obey the traditions to the degree that we need to, while ignoring conventional thought about music and sound design enough to let us do new and interesting things.

I *Have you noticed a shift? Do you feel like there's more fluidity and more interchangeability and more people who are using sounds in their music more recently?*

RT Yes, there definitely has been. There was very little of that sort of thing 40 or 50 years ago, and it's gradually increased. And the degree that it happens has a lot to do with the ability of Composers and Sound Designers to actually collaborate with each other.

Often, we start out with the best of intentions about collaborating, but each department eventually gets swamped with work. And you really have to put your blinders on at some point and concentrate on your own work. So often, those good intentions about collaborating and trading material back and forth between music and sound design get lost somewhere along the way.

But it's also a function of how interested the director is in having those departments collaborate. Some directors are very interested in it and very serious about making sure that it happens, and others aren't. Far too many directors just cross their fingers and hope that somehow the soundtrack will all turn out well in the end. But it's odd that they rarely have that kind of attitude about on set collaboration. They make very certain that the production designer and the cinematographer are talking from early in pre-production about the colour palette, shapes and composition, costumes, etc. But we have this odd disconnect where similar discussions don't happen about sound design and music.

The Potential of Sonic Storytelling

I   *I know you've been very vocal about the benefits of sound being involved in the script stage even and being there at the beginning, and a few other people we've talked to have said that their greatest collaborations are with directors who invite them in at the beginning of the process to allow sound to do its thing.*

RT   Yes, as I've said many times, until fairly recent years most directors didn't have much training in how to use sound. In most film schools sound was treated, and still is to some degree, as a kind of boring series of technical exercises that you unfortunately have to go through to get the film finished and released, as opposed to sound being an active collaborator in the process. To be an active collaborator, you really have to be involved in the beginning.

      Another challenge is that most Screenwriters don't think a lot about sound. And unless they've worked with a Director for a long time, they're usually reluctant to clutter their scripts with descriptions of sounds because the rule is that when you're writing a speculative script (so-called spec script), or a script for somebody who you haven't worked with before, what you want is really just the bones of what's going on and the dialogue. Because it'll be considered presumptuous for the writer to put in too many details about what things might look like or what things sound like.

      But I'm always encouraging scriptwriters to please think about how sound is going to function in the story, and how the story and the production will facilitate that function, while they are in the process of writing the script. Because if you don't, it's very likely that a lot of the good ideas that might be out there to be grabbed will never be found.

      I've worked on so many films with very bright, wonderful creative directors to whom it just didn't occur to use sounds in ways that would have made the film better. And the way to implement that kind of sound is very often by setting it up visually and in terms of dialogue and in terms of pacing, in ways that you can only do when you're shooting the film, or before you're shooting the film, when you're rehearsing the film. Some of these ideas just cannot be pasted on in Post-Production.

      So that's a tough nut to crack because there's a lot of inertia in the traditional filmmaking practices that goes against that idea of thinking seriously and deeply about sound, and how you're going to set up the film to use sound.

I   *Are there any ways or approaches that you've used in any of the lectures that you've given to try and encourage students to think about sound in a non-technical way?*

RT   One of the things I recommend is to show them the first five minutes of *Once Upon a Time in the West*.[16] I mentioned *Once Upon a Time in America* a while ago, but I think *Once Upon a Time in the West* may be, in some ways, the most profound use of sound in film history, in the sense that sound determined how Leone[17] shot that sequence in the film.

      One of Leone's ideas for that film was that Ennio Morricone[18] as Composer would write an entire score for the film *before* they started shooting. And Leone was going to then play back the score on the set to help inspire the actors and just to help shape the making of the film because he had so much respect for Morricone. So

Morricone did that. He wrote a score for the whole film before they started shooting, and recorded it, but neither he nor Leone was happy with the first cue, the first piece of music in the very beginning of the film. And they didn't know what they were going to do about that.

But finally, before it was too late, Morricone happened to go to a Music Concréte concert where a guy was playing a ladder, scraping and pounding on the ladder. And a little light bulb went off in his head and he called Sergio Leone afterward and said, "There shouldn't be any conventional music in the beginning of the film. The music in the beginning of the film should be the sounds that happen in and around this little train station, where these three very bad guys are waiting to kill somebody who's arriving on the train."

And so that's what Leone did, he shot the entire opening sequence using sounds as his guide posts. The very rhythmic and harmonic squeaking of the windmill, the dripping of the water, etc. So why shouldn't that happen more often? It's kind of incredible to me that it hasn't occurred to more high-profile filmmakers to shoot sequences like that, in which the actual sounds in a location drive the narrative.

I think that's the epitome of what I would like to happen, for directors to think about sound in that way, and consider it important enough to use it as a guide.

Proximity as Fundamental to Drama

I   *I wanted to come back to this question of emotion because there's the danger that lots of discussions in sound become very technical – we talk about techniques and approaches – while the actual power in sound often lies in the emotional potential that it has, and the direct emotional communication that it can have with us. We feel sounds.*

A few people we've spoken to have talked about texture and the texture of sounds. And others have talked about the motion and the gesture of sounds as being the driver of these emotional relationships. Do either of those notions resonate with you in the way that you think about how sounds communicate with people?

RT   Sure, yes, all of that resonates with me. Personally, I think that the most drama is to be had in how close a sound is to the listener.

The proximity of a sound relative to other sounds or the changing proximity over time can make a huge difference. If you hear a lion growling that sounds like it's 100 feet from you, that's a very different emotional experience than hearing a lion growling that's three feet from you [laughter]. So, I think that dimension has the most power in it, in terms of playing with sound, how close a sound is to you.

After proximity, the left/right dimension is fairly trivial because it doesn't really matter whether the lion is three feet from you on your right or three feet from you on your left. It may be interesting, but there's essentially no difference in drama there.

Likewise, it doesn't matter that much whether the lion that's 100 feet from you is in front of you or behind you, or below you or above you.

All of that's to say, I think we sound nerds tend to spend far too much time thinking about spatiality in the left, right, front, back, top, and bottom dimensions than we should. And we should spend much more time thinking about proximity.

And that's an element of sound that very much should inform directors when they are setting up shots, thinking about that element of sound. And getting away from lions and getting to people, there's a huge difference whether somebody is whispering who's six or eight feet from you, and somebody who's whispering when they're a foot from you. And so I wish directors would play more with that kind of dynamic in terms of sound. But yes, sounds have textures and sounds move. And all of those things affect us emotionally. They're all emotional vectors.

I   *Yes, yes. In a sense then, it's about this balance of shifting perspectives of sound and bringing certain things to the fore and pulling other things back, and through that change and shift that charts the pathway for the listener through those sounds.*

Capturing Sounds

I   *Do you value the process of going out to record your own sounds?*

RT   When I was much younger, I would carry a recorder of some kind with me almost everywhere that I went. And there's enormous value to be had in collecting one's own sounds, partly that you always discover sounds that you wouldn't have anticipated being in a given place. And it's almost always a sound that surprises you that turns out to be the most valuable sound that you find in a given location. I was working on a film fairly early in my career that had a nuclear power plant in it and so I went to a nuclear power plant to record. I didn't have much of a notion of what I would hear there, but I imagined mysterious sinister sounds of all kinds. And it was one of the most sonically boring places I've ever been. It's all just fans and humming and nothing sinister sounding at all. But the water fountain, the drinking water fountain outside the manager's office was malfunctioning, and that's where the sinister sound came from that I wound up using as one of the elements in the nuclear power generation system [laughter].

If you are a professional in sound, certainly sound effects, I think, eventually you will use almost every sound you ever record. You may not use it for the project that you're working on that day, but you'll use it somewhere down the line. So I'm always encouraging, especially young sound recordists, to just record all the time, partly because recording media is cheap these days. Just go to a location and let your recorder run most of the time that you're there. And I guarantee that there will be a sound or a set of sounds that will happen that you wouldn't have anticipated, that'll be enormously useful to you at some point. All that said, I think any creative sound person can find enormous value in almost any sound effects library that's already been recorded. Obviously, you don't want to use exactly the same sound in exactly the same way that lots of other people have used before, unless you're doing it as a joke. And most of those jokes get fairly old fairly fast.

But if you take, let's say, a fairly well-known sound effect that has been used in quite a few films, but you use it in an entirely different way, associated with completely different visual images than it has been in the past, then it's for all practical purposes, a new sound because it'll be perceived as a new sound. And nobody except the extreme sound nerds will even recognise the sound. They'll not even know that they've ever heard the sound before because context is everything. And so if they're hearing it in the context of a different set of visual images, it'll seem like a new sound.

So, there's that on the one hand, and there's also the fact that our ability to manipulate sounds is so profound these days that you can take even a fairly familiar sound and modulate it or manipulate it in ways that will retain the essential emotional value of the sound, but make it pretty unrecognisable for people who may have heard the original version of that sound before. So, I try to approach all sounds as raw material to be played with and manipulated and experimented with in all kinds of contexts.

Ren Klyce[19] is a contemporary Sound Designer, one of the best; he does the sound for David Fincher's[20] films among others, and one of the things that I admire about Ren is that he tries to record every sound for every project that he works on new. So he tries his best to avoid ever using a library. And that's a daunting task. And so I admire him for it. At the same time, I think it's probably not always aesthetically the best choice because I think sometimes you could find a sound in a library that would be more powerful or more useful than one that you happen to have been able to record during the last few weeks or a couple of months that you're working on the project. So, while I admire the discipline of it and the philosophy behind it, I think my approach is to use a combination of trying to collect as many new sounds as I can or ask the people who are working with me to collect as many new sounds as possible for each project, but not to rule out looking through libraries.

I   *It seems like you're talking about a process of discovery, a way of seeking out the unexpected, either through recording or by exploring sound libraries?*

RT   And as I probably said, one of the first things that I do when I start working on a film is just randomly listen to sounds in a library. Having read the script, or maybe seen some footage, I then just start randomly listening to sounds without any preconception, without any attempt to think, "Well, this film has lots of cars in it, so I'll listen to car sounds".

I literally just start clicking randomly through the library, and inevitably, within a few minutes, I find a sound that I never would have anticipated. It never would have occurred to me to put it on a list of sounds to look for any of the scenes in the film, but it's a sound that has really interesting profound connections or a connection of some kind, to something that's in the script or some bit of action. And I think it'll be powerfully metaphorical or enhance some more obvious sound that I might find later to use in that sequence. I think that a John Cage[21] style, chance operation approach for Film Sound can be really, really useful.

The Role of the Unexpected – Process/Chance/Experimentation

I   *It's interesting, the question of the unexpected is something that's recurred through our discussion today. You talk about it in terms of providing something unexpected for the audience in order to stimulate their sonic creativity, how the unexpected sounds that you find when you're out field recording are often the most useful, and this process of chance listening to library sounds to break out of obvious initial choices.*

Do you think that this question of the unexpected and the chance is a core part of your practice?

RT   Yes, it absolutely is. None of us are smart enough to invent a world. So inevitably, we invent worlds by selecting pieces of existing worlds, and combining them in a way that will make it seem like a new world. I think almost all art is really editing. Whether you're a visual artist or a sound artist or a writer, the most crucial thing that you do is edit. You expose yourself to as many ideas and notions as you can about whatever project it is that you're working on. And then you set about deciding which of those things are going to be useful and how you're going to combine them. So that's what I'm calling editing. And very often, the unanticipated element is the one that proves to be most useful or most interesting, most profound.

I often say: A great craftsman is somebody who knows how to avoid accidents. And a great artist is somebody who knows how to use accidents. And I think that really speaks to the most important part of creativity as far as I'm concerned. It's the accident.

People who don't know much about art think that the artist has this grand vision of what the piece of art is going to be, and then actually doing the art is just this series of perfunctory operations, using the techniques that the artist has assembled over the years, but that's really not how most art happens at all. There sometimes is a grand notion, but the grand notion almost always gets altered in important ways in the process, and it's the process itself where the most interesting stuff happens. It's the discoveries that the artist makes, painting the picture, or doing the sound work, out in the field with a microphone, just listening. It's those discoveries that are, I think, most often what winds up driving the piece of art.

So, the mindset that you go into it with is so important because you have to maintain this balance between having read the script and having talked with the director about how they see the project; you have to keep all that in mind, but at the same time, you have to keep your mind open to the unanticipated, the things that don't have any obvious connection to what the director has told you or what's in the script, but are things, in fact, that can be made to be very connected to both. So, I think that's the creative mindset that I always try to be in. But it's not always possible.

I   *Yes, it's fascinating isn't it, that often people think, "I must have a master plan and then set out to realise that master plan", without recognising that most of what we do is a discovery, a process of finding out through doing, that is only accessible through doing. And until you start working with those materials and those processes you don't really know what the possibilities are or where you are going to end up.*

RT   Yes. And once again, I think it has something to do with complexity. Even a relatively complex thought is simple compared to actually doing something in the real world, where all kinds of variables enter that would never have occurred to you to anticipate. So when you find yourself there actually manipulating the sounds, it's just inevitable that things will happen that wouldn't have occurred to you otherwise. So, the trick is to figure out which of those things will be useful to you and which won't. And you never know until you try, until you play with it for a while and make some mistakes.

That's the other important element of all of this. You have to have time to make some mistakes if you want to do anything very interesting, because if you don't have time to make any mistakes, then you're just going to duplicate something you've done before.

I   *And I suppose it's also about being confident enough to allow yourself to make mistakes, and to try things out. This is something that people are sometimes afraid of discussing?*

RT   Yes, there is a real pressure to put up this façade, "I'm a professional. I don't make mistakes. I know exactly what I'm doing at all times". You have to pretend that is the case because that's often why the producer or the director hired you, because they think you already know how to do whatever it is they want you to do. But secretly you know that you don't really know how you're going to accomplish whatever it is. And even when you've been doing the work as long as I have, I still have mild panics often about how I'm going to come up with a sound or a set of sounds. And so, I just need to teach myself to breathe deeply a few times and try a few more things.

I   *Yes, I'm really fascinated by this question of chance and this process of discovery as a form of practice.*

RT   I think what we call creativity is more often actually discovery. You're not really creating something; what you're doing when you get lucky and you work at it hard enough is discovering something that in a sense is already there, but maybe not there in the order that you need it to be in. So, it's about finding these elements that are already there, and which you can rearrange in a way that they'll be useful.

4.1.4 Nina Hartstone

Nina Hartstone (b.1971) is a Supervising Sound Editor, specialising in Dialogue and ADR. The first European woman to win an Oscar for sound editing, Nina's work on *Bohemian Rhapsody* also secured wins for sound editing at BAFTA, AMPS, and MPSE. With credits including *Gravity* (2013), *Everest* (2015), *Enola Holmes* (2020), and *Moonage Daydream* (2022), she has worked alongside leading industry professionals to implement innovative ways of recording, creating, and editing sound for film.

In 2022, Hartstone earned the Lifetime Achievement Award at the Septimius Awards in Amsterdam, the Netherlands.

Speaker Key

I   Interviewer
NH   Nina Hartstone

Listening, Creativity, and Performance

I   *Is there a reason why you've been drawn to ADR?*

NH   I came into my career in sound learning from the bottom up. I don't have a grounding in the technology or the physics. A lot of my colleagues have a really geeky fascination with the science of it, and I'm more interested in the listening and the creative side. That's where my experience comes into play, from the years and years of listening. I can use all the technology and make it do whatever I need it to do, but I don't get excited because there's a new plug-in. I'm led by my creative goal, "I want to do something with a voice, so how do I get there?".

Most women in sound work in the dialogue side such as ADR, rather than SFX. And there's a bit of sexism in there because ADR involves typing, paperwork, and organisation, and it involves an emotional connection with the actors that can be hard, those touchy-feely elements. It's dialogue, it's drama, and it's emotion, whereas 'bangy crashy' sounds are supposed to be quite male.

When I was an assistant, most of the work was on the dialogue side because there was so much prep to do before the recording sessions – working with the script and finding alternate takes etc.. So, it was natural to take jobs in dialogue.

Now I'm a supervisor, I feel like I'm overseeing so much of the soundtrack, and I'm quite happy not to be sitting there making sounds, creating sound design in that way. I love the fact that I work with the actors, and I get to record voices in the way that I do, because when I was a teenager, my aspiration was to be a director, and so, working with the voice feels more a natural path for me.

The backbone of most of our storytelling work is made in relation to vocal performances on camera. Everything we do is artifice really, so these stories need to *feel* as soon as you start, and the way you get drawn into a story is by not questioning any of it. You need to *feel it*.

For example, the Live Aid sequence in *Bohemian Rhapsody*[22] – where they perform on stage at the end – was a complete creation. Rami [Malek] was pretending to be singing, but it wasn't his voice coming out of his mouth. But the most the overriding factor in that sequence has to be that you never questioned that voice [of Freddie Mercury] would come out of [Rami Malek's] mouth.

Feeling and Understanding – Technique and Process

I   *I guess that's the challenge of the practice, that transparency, if it's doing its job, then you don't draw attention to the detailed work that has gone into making it. But at the same time, you are highlighting all of these small subtle nuances and details in the voice and creating a sense of proximity with the characters.*

NH Yeah, it draws you in. I really like the sound that you can capture with radio mics because of the detail you get. It's almost as if you're right inside that person's mouth. One of the things I focus on within my ADR work is really trying to capture all of those tiny details, to anchor whatever I've recorded to the mouth.

I   *And you record extra takes where you just capture textured breaths and vocal gestures?*

NH Lip smacks and breaths, yes. You always want to try and capture it in one take, as a flow if possible, but it depends very much on the actor. I'm not beyond asking an actor to do separate passes of the scene with just breath, just lip smacks, etc. I'll then use all of these individual elements to try and recreate what we had originally.

   The most important thing is to make it feel real. The stuff that is most important is all those 'soft' bits. If it's not properly fitted[23] and the detail gets missed, then that will just put up a barrier between the actor's performance and the audience.

   With ADR for example, there's often quite a temptation to cut out the bits between the syllables, but the reality is that nobody can speak like that. You just can't. Speech has to have moments of pause to accommodate the physical movement of the mouth, and pauses are essential for you to create that next sound. You can hear if something has been cut in a way that is contradictory to the natural process of human speech because it sounds robotic.

   My hearing has become very attuned to those subtle details over the years, and if you're trying to cut down a take to make a line fit, sometimes you're often better off taking the cut out of the sound rather than the air in between. In one line, it might not matter, but across the whole journey, it'll just push you away rather than pull you in.

I   *I suppose the voice is probably where you are most attuned as a listener, you will feel if it is mechanical, or if the levels are a bit out of balance?*

NH Yeah, absolutely. I often get asked lots and lots of technical questions, for example, "what levels are you using when editing"? But my reply is always, "I've just listened to it". If somebody is speaking, there's a natural level that you'd hear it in the real world, if they're whispering it's going to be quiet, if they are shouting it's going to be louder. It's just the way it is [laughter].

   I feel you can really tell where the right spot is level wise, you can tell if they've been pushed, and you can tell if they've been held back. So, I tend to use my ears to gauge, rather than keeping an eye on the [VU] meter[24] trying to ensure that all of my dialogue sits within a certain dynamic band.

   Particularly with dialogue, you understand what feels right, or what makes you go off kilter. It's a little bit like looking at a visual effects sequence that doesn't sit right with your eyes.

   I'm always trying things out to see what works. When we've got ADR I'll put a line on it and tweak it if something is bumping me out of the story when watching. I just keep trying different things until I discover what works.

Authenticity = Emotional Connection

I   *I'm interested in this question of authenticity. This balance of the 'real' versus the constructed and how you deal with this balance. Because you're constructing things that are not real, but you are basing them on real principles, how do you navigate that?*

NH   As I said before, you're trying to establish an emotional connection between your viewer and their story that they're watching. So, you use your reality to draw them in, but of course, everything we do is artifice.

   If we have an exterior crowd scene, I always like to do the recording of the crowd group outside, because it feels more real.[25] But we do also cheat a lot. For example, for the Live Aid scenes in *Bohemian Rhapsody*, we had only a certain number of people singing from which we had to create the impression of a whole stadium. If you're telling something realistic, you want to make it feel as real as possible and you want to do everything you can to try and make it *feel* real.

Layering and Sonic Focus

When working on a crowd scene, we obviously re-create the voices for the whole crowd, but I always find that in any sequence my eyes are always caught by the same few, maybe three, people. And it might not even be that you see their mouth, it might be someone who's punching the air, it might be someone leaping, any movement is what generally catches your attention. So, it's about articulating those moments that have drawn your attention.

I   *Walter Murch talks about his rule of 'two and a half' where he says there's a delicate balance between layering individual sounds and conveying masses. If you've got two people's footsteps, you very clearly hear them as two individuals, but if you have three people's footsteps, they all of a sudden appear as a group. Beyond three he suggests is redundant because if you have too many layers it can become too complicated and messy, so he always argued for trying to convey the impression of masses with as few ingredients as possible. Do you find that this issue of density is something you come up against?*

NH   If we take the example of vocal groups,[26] a lot depends on how you capture or mic them. Recording in these times of COVID,[27] we've been discovering some interesting things. I always like to try and keep things separate because I find that recordings of groups into one mic quickly become a mush. But with COVID safe recording, each person has their own individual mics and we immediately noticed that when you place those stems together, it's not a mush anymore.

   So, it's more about having that separation, and how you've recorded it, and I wonder if you had distinct recordings for each of those people. And if you were able to play with the level of them a little bit, to place them at slightly different distances, pan them across the front speakers? I wonder if you could be more exact about it today?

   For *Bohemian Rhapsody*, we had a marketing app where they got members of the public to sing. It was called *"Put me in Bohemian!,"*[28] did you know about that?

I   *No, I didn't!*

Subtlety, Variation, and Richness in Voice Textures

NH   They released an app that encouraged people to use their smartphone to record themselves singing along to *Bohemian Rhapsody*[29] and submit it to be included in the film. So, we received these packages of thousands, thousands of recordings, of people warbling along to *Bohemian Rhapsody* [laughter], which we had to line up and mix down, and line up and mix down, etc. But having them all as individual recordings meant that it sounded very different.

We went to record crowds in the O$_2$ arena, but somehow these sounds from the app sounded even bigger, which I found really fascinating. Each of the individual voices still held their subtle detail and differentiation, even after you've mixed down thousands and thousands of voices.

When they first mentioned this app, I thought "that's probably not going to be terribly useful", but you know what!? It was *really* useful! [laughter].

Recording and Acoustic Space

I   People can often demarcate in their head the process of capturing the sound and the process of editing the sound. And I just wondered if you could talk a bit about how important that process of recording the sounds is to your whole working method?

NH It's totally crucial really, because the recordings that we get are our source material. And you are limited by the source that you have. There are amazing things that you can do with different plug-ins: adding reverb, changing EQ, and changing levels. But if it isn't on the recording in the first place – if you don't have the frequencies that you want to boost, or you don't have enough signal-to-noise ratio – you can't do anything about it. Even adding reverb via plug-in doesn't beat an actual recording, in a real setting, where the microphone is a set distance from the sound source and you have the real feel of it.

Everything that's done by algorithm has a form of one size fits all. And I know so much of the work that goes into plug-ins is trying to randomise things and trying to make it feel real. But it's always an imitation of what happens to a sound in the real world.

So, that's why we're always trying to find the right sound first. If it's for sound design, make sure that something about it is already right to the picture, whether it's the rhythm, whether it's the frequency, whether it's how aggressive or soft it is, those things have to be there in the initial sound. Then we can add or enhance. We can pitch things down, push things up, all sorts. But the meat and potatoes need to be there at the beginning. If we don't have that from the get-go, it's very difficult to create something.

You can't take something and transform it into something completely different without noticing the process that's taking you there. And this is especially important because what we do is about believing the sound that you're hearing.

You never want to hear the process that a sound has gone through. You want to believe that that sound you're hearing is exactly as it was, if you'd heard it in a real-life environment rather than through the speakers.

Sonic Clarity

I   There is something in this idea of randomness and chance and the unexpected. One challenge we face is that the tools and technology we have are often designed to help you to make things super clean. But, maybe that's not the best creative move?

NH For sure. We don't usually work with the idea that it has to be super clean. If I'm working on the dialogue, I'm only one piece of the puzzle, one component of the frequencies that are going to be heard in the mix. If you play one sound on its own, it will sound very different than if you layer it up with other sounds.

So, if you over process a sound because you want it to be completely clean and pure on its own, that's one thing, but if it's not likely to be heard on its own in the final mix, then you're not actually working towards the right end. You can be doing too much treatment to sounds that can introduce artefacts and artificiality, and you risk impacting on the connection that this dialogue has with the picture.

It's really important to be aware of context and listen to how it's going to sound even with the temp music and temp FX. How thick are the atmospheres going to be? How much Foley are we going to need? How much cloth is there going to be? Often, you get a scene where someone is running down the street and you've got loads of ambience going along, plenty of dirt in there, you might be hearing their feet and their clothes and everything, and as long as the noise of the voice, the signal, is loud enough to rise above all the other stuff, it's better to leave it in really, that extraneous sound. Don't try and make it just pure voice because instantly you just remove every connection it has with the image that it's been recorded with.

Actually having the roundness of those 'noisy' frequencies in your source material will often allow it to bed in, rather than float on top of the other sounds. It's often all of that extraneous sound that you might think you don't want, which is the thing that glues all of our sounds together. Pure sounds sound like distinct layers, and they will never combine into one melded, rendered picture.

You want everything to mush into each other, to lock in rather than just sit alongside. And it is often the 'air' around the sounds that helps bring it all together and smoosh it in – that's not a very technical term I know – but you want it to all smoosh together [laughter].

Sound Relationships/Reality – Rhythm

I   *I'm interested in this notion that you don't just have sounds, but you always have sounds in a space, they're always in relation to something else.*

NH Very much so. We don't tend to work on sound in isolation; they're always in relation to the images, that we're seeing, but also the story the whole piece is trying to tell. It's not just a visceral experience of listening to sounds, it's the experience of the entire piece of media that we're working towards.

So much of the time we are trying to create a reality, and the sound is there to transport the audience into what you're seeing. Even if the characters are up in spaceships and surrounded by unfamiliar sounds, you still want the audience to feel anchored in reality.

You're trying to get your viewer to empathise with your protagonist, so you want the viewer to feel like they are that protagonist, and that they are in that space, that they can hear the space around them.

But, when you get subjective and go into someone's head, all bets are off! You can go down a completely different journey, you're trying to convey an emotion rather than set the scene that you're visually seeing. For example, in *Fantasia*,[30] we were working much more with the rhythms of the images and the emotion that the images are trying to convey. You're working much more with colour and speed and shape and rhythm, particularly the rhythm.

Rhythm Across the Edit – Example: *Gravity*

I   *I'm really interested in this notion of rhythm. It makes me think of a comment that you made about your work on 'Gravity,'[31] working rhythmically with the patterns and the textures of the breathing.*

NH   Yes, the breathing was a huge aspect of that soundtrack because your connection with her character was literally through her breath. Even when she's spinning away, far off into space as a tiny dot in the distance, Alfonzo [Cuarón] always wants to feel as if you're very closely connected with her.

When we recorded [Sandra Bullock] and George [Clooney] in Post-Production, they had three mics: one on their forehead, the second over the ear to mimic the NASA costume mic (which would catch all of those mic blows [breaths]), and a boom mic.

The biggest challenge was trying to cut it together so she never stops breathing normally. In most other films, you can interrupt the rhythm of something by cutting away. But in *Gravity*, it became quite a mission to create breathing that works all the time. Every single piece had to follow on 'naturally' – an inhale and exhale, inhale and exhale, or a hold – so her transition from faster breathing to slow breathing really made me work.

It also made me realise that breathing is not just inhale/exhale, it's not just nasal/mouth, there is so much more to it. I almost want to sit and document the different types of breaths that there are!

You can see from the shapes of the waveforms, the ones that have more power at the front, ones where you breathe out and you hold, or they breathe in and hold. It's also present in the editing, where you leave it. Whether you have the breath on the end of a line of dialogue or not, or whether you leave it hanging? How sharp it is at the front, or if it builds up? If you end with an inhale, it creates a sense of tension, while leaving with an exhale feels like everything's settled in that length of the breath. All of those different elements are the things that I'll work with as I try and build the emotion of the character.

All those details you can really hear; even though, ostensibly, it's something where you could say "well they're holding so you could just cut it", but it doesn't work; there's some texture in the air when you hold your breath that you hear.

Morphologies of the Voice

I   *I suppose it's a result of the muscles straining to hold the position of the lungs?*

NH   Yes perhaps, but if it's not there, it doesn't feel right. There were certain sequences that got quite tricky. Quick transitions between calm and panic, back to calm and then panic again! And it got quite difficult to find the right rhythm that matched visually. You set up a match every time you can see her face, but she might turn and "oh now she looks like she's breathing at a different pace", so you have to think "how do I transition between these different rates as we come back round to her?" It needs to feel organic. So, sculpting all of that became quite fiddly [laughter].

But [Sandra Bullock] did an amazing job, most of it was ADR by the end because it needed to be. They were in these contraptions that created a horrible [servo noise] every time it moved them into a new position. So, we recreated all of it in post really.

Emotional Honesty and Physiological Detail

I   *How do you judge what is right and wrong, and what feels real?*

NH   I spend so much of my time listening to voices and conversations and pulling them apart, listening to the tiny, minute details. For example, I'm really attuned to subtleties in the voice like the type of breath that precedes somebody crying. Not the breath they have when they're crying, but before they're going cry. These are very distinct sounds.

I often listen to ADR takes and say to myself, "they're not smiling enough". It's not because there's any laughter on the voice, it's literally just the way that the face shape changes as the voice comes out. Somehow you can hear it in the voice.

Quite often the actors are very focused on the technical details in ADR. So, I have to remind them "you look very relaxed and you're smiling in this one and I can hear that in your voice". I'm not sure if that nuance is something that most people listen out for but that's something that I've built into my practice over years and years of working with this material.

I   *When you're talking about listening, and you're training yourself to listen are you always talking about film clips? Or do you do a lot of listening out and about, for example, when you're at a dinner party?*

NH   Particularly when I'm knee deep in a project, I feel hyper aware of all voices, and I'm almost cutting out the ones that don't like. I'm literally editing things in my head as I sit in a restaurant [laughter]. You're living and breathing the job, and the relationship that you have with sounds is so close. And inevitably that's how your head is thinking about sound, so when you turn away from the computer, it hasn't gone away, you are still listening in the same way.

Teamwork, Subjectivity, and Diversity of Perspectives

I   *To what extent do you use your experience of listening in the outside world to influence the decision-making process that you have in the edit suite?*

NH   It influences it quite a lot really. Everyone experiences the world through their own eyes and ears and we can only have our own impression of things. This is why I really love having a team and having fresh ears, because it's so valuable to try and get more than just my own subjective opinion on what does and doesn't work.

You can convince yourself that something sounds okay in the edit, but often I'll look at it the next day and realise, "Oh, my God, it's terrible" [laughter]. Because it's all so comparative to where you were a moment ago. I can only bring *my* own perspective to the work, so that's why I do love to hear other people's perspectives.

I think that you get the best out of your team by trying to make sure that everybody feels heard. I want to hear everyone's opinion, it's nice to have that breadth of things, and you definitely get more creativity if your crew have a broader base of experience, than if you just have one type of person creating a film. You want your film to appeal to everybody in an ideal world. Your viewers are a broad range of people, so it's important to bring that alternative experience.

Otherwise, you're missing a reach on certain aspects of the story that you could highlight, and I'm not just talking about overarching narratives but also subtle elements, all those little subtleties build to add extra dimensions to characters.

If everyone's got the same perspective, it becomes very, very boring. There are fewer creative discussions. So, I don't think it's best for your final output.

I   *You've mentioned collaboration and embracing multiple voices and trusting other people's opinion; letting the fact that you're there as a team be the valuable thing and not just ensuring that the team is enacting your own vision, it's fundamentally about how you can make this thing together?*

NH   It is interesting to be challenged, I want to hear other people's opinions, and I don't want everyone to agree with me. I might not have thought of it every option, and it's really useful for people to give me another perspective so I can consider it.

People find it difficult to step into someone else's shoes, and the only way you can understand that is by listening to the experience of other people and understanding what life is like for them, where they are, and how they experience situations. The more we can learn about everyone else's experiences, the better it is for everybody else.

I   *And it strikes me that maybe part of this is also about empowering people to feel confident to express their opinions.*

NH   It's even something that I face. I'm very fortunate in that I usually work with fantastic teams, and it doesn't even occur to me whether I should open my mouth or not. But this is not always the case. Sometimes, I sit in the room and worry about whether I should express an option, and it's very clear that this doesn't even occur to some of the men in the room.

It would be good for everyone to understand and think about the experience of others a bit more.

It's also important to think about the story and overarching narratives in the film. For example, in *Bohemian Rhapsody*, when all the actors first come on stage at the Live Aid gig, we go into a little slow-mo bit with Freddie, and we cut to each of the band members. They're all very anxious, and they're breathing very heavily. In my dialogue head, I'm thinking, "oh, we need all of their breathing so you feel their anxiety". But, when I wheel myself back and look at that sequence in context, I felt differently, because you have to see things as a journey.

In some ways, I would love to hear them breathing, but it's less important to me than the feel we want in the wider sequence. If we hear them breathing here it's going make it feel smaller and focus us down onto tiny details, but we're in a large-scale event, so having silence here contributed to that sense of scale.

Giving sound to each little moment was not going to help us in terms of telling the story, and that's how it ended up in the final edit. It's one of those things where I have to balance my dialogue head with my sound supervision head [laughter].

Your end goal is to get your viewer to be subsumed by the story and not think about anything else, as if you're diving into a good book where you're not aware of your surroundings, because you're so into what's in the story.

I   *That shift of perspective is so important. I love this idea that the different parts of your creative brain are competing; one part is saying, "Oh, we must edit in the small sounds here". But then, the other part is saying, "Well, maybe not". It's that creative decision process in action. If you have those tiny little sounds, they draw you into the small detail and make the scene feel more intimate. But that scene is not intimate, they're on a stage in a massive stadium, and you want to encourage that sense of scale and work with the trajectory of the larger sequence.*

Emotion, Embodiment, and Sonic Gestures

NH   It's similar to when you record crowd artists crying, or scared, or in pain. What you realise is that when somebody is experiencing heavy emotions, they're actually breathing less. What they're doing, most of the time, in those situations is holding their breath.

When you experience a strong emotion, whether it's pain or upset or fear, everything in your body is trying to just hold, and so the sounds aren't loose, they're escaping.

It's tempting for people to be melodramatic when they perform that type of stuff, but if you see someone genuinely upset, you'll see that most of the time they're trying to manage the sensations inside their body at that moment. And that doesn't sound the same as letting it loose continuously.

Plus there's more drama told in the held breath [pause] before the next bit bubbles out.

I   *That makes perfect sense, the tension that you get into when you're in shock and how, of course, the voice is part of the whole body. It's not just your throat, it's about your lungs and your back, and even what your arms are doing. It all changes and modulates the sound that comes out. When you're working with actors in the studio, do you talk to them about their whole body?*

NH   Yes, we're very physical when we do ADR, so I always have a boom op in as well. Standing in front of a lectern talking into a microphone just never feels right for the actors, and it never feels right for the recordings you get. So, I will get them to do whatever helps them be that character again in that moment. It's really about being as physical as you can.

In ADR, we're always saying 'cheat up the movement', which is to say, "you're not doing much on screen, but you're outdoors, you've been walking for a while, generally you just breathe a bit heavier when you're out and about".

If you fake that it just sounds like you're pushing air from the top of your chest. It's not genuine deep breathing caused by your cardiovascular system being elevated. So, I get them to bounce heavily so that you can hear the movement on the voice as it's bouncing up and down. You need them to exacerbate that in ADR to be able to fit it on someone who's even not moving that much.

On the film *Everest*,[32] they were throwing themselves around. We get people moving all over the place in the studio, they can run, they can do push-ups, they can be lying down when they perform, and they even had a sequence in *Mowgli*[33] where they had the actor hanging upside down, leaning over a chair, to get his upside-down voice. Those details also help the actors because suddenly it all feels more real. It also helps me hugely, because the recordings that you get don't need to be perfectly clean, they just need to sound right for what they were doing visually at the time.

Directing the Performance in ADR

I   *I suppose it's this idea of not just extracting the voice from its context. All sounds really are always within a context when they're recorded. And that is something you are sensitive to all the way through your practice, whether it's recording sounds in a real space with real reverb or retaining the noisy bits so that when you 'smoosh' the layers together they gel. And in the performance with actors as well, there is always that sense of the bigger picture.*

NH You're quite right. It's everything. And there's an awful lot going on in their heads that needs to come as well. We've had actors with all sorts of different processes, some that get a bit cross,[34] but you just let them do it. Because sometimes they're doing that to feel some emotion before they do their lines. So, it's better to just let them have their process and get on with it. Because it hopefully gets them where they need to be.

I   *It's probably one of the most people-centred aspects of the whole Post-Production process?*

NH It's akin to working with the actors on set. But my role can vary greatly in the ADR theatre, if the Director just wants me to be there for the purely technical process of recording, that's totally fine. Other directors will say "You know what, I've got loads of other stuff to do, take the session on your own", or "I'll be here, I'm doing stuff and you get me to listen properly when you've got something that's mostly there".

But there's such a strong overlap between the technical and the performance, I can be giving technical notes like, "You need to speak more softly", or "You need to drop off at the end of the sentence", and through these, you really are starting to talk about performance.

I   *I suppose the fascinating thing about it is that you really can't pull the two apart, it's inherently in the storytelling, the way that you articulate sounds is a storytelling act.*

NH One hundred percent, it totally is.

4.1.5 George Vlad

George Vlad (b.1985) is a Sound Designer, Sound Recordist, and Composer. With a passion for recoding sound environments that are largely untouched by western society, he has embarked on expeditions that have taken him all over the world, from the Jungles of Borneo to the Kalahari Desert, the savanna of the Masai Mara and the Amazon Rainforest. He creates audio for video games through *game-sounds.com* and captures studio and field recordings for *Mindful Audio*. George also travels to remote areas of the world to capture rare soundscapes and releases field recording albums through *Wild Aesthesia*.

Speaker Key

I	Interviewer
GV	George Vlad

Sonic Discovery

I   *Do you remember when you first discovered this world of playing with sound?*

GV   I grew up with my grandparents in the countryside in Romania, I remember they had a huge metal pan that they used for cooking at big gatherings, and it had this musical resonance. As a child, around four or five, I spent hours tapping on it with various objects and trying to find the sweet spot, where it made the most pleasing sound. And people would just look at me and ask, "Why are you doing that, just go and play with the other kids instead of sitting there and banging on pots" [laughter].

I was never aware that this was actually a 'thing'. I just did it because I liked it. There were many sorts of anecdotes like these during my childhood, and into my adulthood as well. When I was 13 or 14, a friend bought a computer, and he was using this weird software to create sound and music, and I got hooked on that. I went on and did some DJ-ing for a few years, though I was never happy with the music that I could play. So, I ended up composing my own, and through this process, I realised that I was most interested in sound design, creating patches for synthesisers, and somehow manipulating these sounds, and creating a sense of space using effects.

I   *It seems that for you, this process of listening is at the real heart of your practice and that sensitivity to sound?*

GV   I think a lot of my memories from early childhood are mostly sound focused instead of being visual. And when I compare my recollection of events with other people's that we experienced together, I do tend to focus mostly on the sound or that some sounds stand out for me whilstvisuals I seem to forget quite easily.

The Potential of Natural Sound

I   *You mentioned technology and developing new skills, and using new tools, what's the appeal of technology?*

GV   There was an episode in my life where I was just interested in electronic music and electronic sounds, and I was not interested in anything natural at all. Everything had to be synthesised or created from scratch. Years later and I've come full circle, now I mostly focus on natural sound!

When starting out, I felt like the natural sound was bounded, with limited variety. Whereas the promise of synthesisers was that you could create virtually anything. But the more I travel, the more I explore and the more I listen, I realise that the world of natural sound is incredibly diverse. I keep discovering new things, new sound sources, new biophony[35] and geophony.[36] And this makes me totally reconsider those boundaries. I now feel that the diversity of natural sound is almost infinite, because it's just so varied.

Whenever someone asks me "What microphones I use", or "Which sound recorder", I get a bit disheartened because the real questions are: "Why did I record that?", "How had I got to that specific place?", and "What my mindset was?".

So, I'm not focused on hardware or technology as much anymore, although I use it as a tool, it's not an end in itself, as it perhaps used to be.

Embracing the Unexpected

I   There is an interesting shift from the known to the unknown in what you're describing, the expected to the unexpected. Working in the studio with the synthesiser, you can create what you can imagine. But with field recording, you can capture things that you can't imagine, because you didn't know that they were going to be there?

GV   Natural sounds can be surprising. You can often discover something completely new that you've never thought existed, for example, the sound of empress cicadas in Borneo at dusk. It's very complex, and the acoustics of the rainforest make it even more exciting.[37]

They sound like something that you might synthesise, but I would never be able to come up with that on my own, it's just too complex to create without having heard it. Now that I have heard it, I'm able to mimic the make-up of the frequency spectrum to create an alien insect that sounds similar. So, experiencing natural sound inspires me, re-enforces my work day to day, and makes me more creative.

Inspiration and Process

I   That's a really exciting way of describing how those two worlds feed into one another, and how that complexity of the world and the chance encounter sometimes exceed your expectations when you're out in the field. I wonder whether we could talk a bit about your process, how you end up in the incredible locations that you do, and the creative choices that you make in terms of how you capture sounds.

GV   My field recording passion started with listening. I had quite a feral upbringing, so even before going to school, when I was five or six, I was free to go roam the hills and forests, and I had friends who had very similar interests. We'd go out, as a group of small children, and we'd map out the surroundings of the village that we grew up in. There was a lot of wilderness back then, much more than there is now.

So, on some level, I think that I am constantly recreating that feeling, gradually enlarging the area that I know around my base. When I started to do field recording properly, I was living in Scotland, and I would drive out into the hills, then into the highlands and then into the national parks, and then I started to fly back to Romania to explore the mountains and other areas I hadn't explored when I was living there. Then, I went to Northern Europe. And then slowly I started to go even further afield.

I went to Africa, South America, and Southeast Asia; and at first, I was just fascinated with exploring a new place and being in a new location and listening to new sounds. I used to record sound primarily for use in games or films. I would visit a rainforest to record and catalogue the sounds, then release them so that other Sound Editors and Sound Designers could use them in their work. That exposed some people to my sound recordings, maybe around a hundred, but it was still a very limited slice of the world and few people could experience them. Then, I realised that a lot of these sounds were endangered (many still are), and I began to feel this responsibility to document and to preserve them. So, that's when I started to share them as albums on Bandcamp.[38]

After a couple of years, I began to see a lot of great responses from people and, without prompting, a lot of the comments mentioned issues of conservation for endangered soundscapes and environments. So, I began to realise that in sharing soundscapes, I could raise the profile of these environments and do something to help reduce their risk, and this is where I started to upload one- or two-hour soundscapes for free on YouTube.[39]

These days, I'm not primarily recording for my own use or for selling sound libraries, I more record for the beauty of it. I like to record people hanging out and chatting in an environment. I like to record deep wilderness camps where there's maybe ten or five people, and you can barely make out the voices among the biophony.

I like to be challenged. And if there is an angle where I can think about conservation and I can record nature in its pristine state, untouched by humans (if that's even a thing), then that draws me in.

I'm not going to expend a lot of effort trying to record the sounds of Southern England, because there are plenty of other people who have access to this place, I'd rather go and record places that might be out of reach for other people. Obviously, I'm mostly limited by how much I can spend to get to places, I self-fund and self-organise my expeditions, and I think I'm privileged to be able to redirect my earnings from game audio to travel on expeditions.

I used to try and get away as far as possible from humans, but eventually I realised that in conservation terms, the human and the natural don't really exist in separate spheres. There will be humans regardless, and there have been humans in these places even though it doesn't seem like it when you're there.

Holistic Soundscapes

I   *I think it's really interesting this shift recently to encapsulate humans as part of the system; one of the great challenges is to recognise that these two worlds are connected, and if you destroy nature, then you are destroying everything for humans as well. It strikes me that perhaps your practice has always been about you and your human perspective on the world? Do you think that's an accurate reflection?*

GV   Not necessarily. My main recording technique is to use something that I call a 'drop rig'.[40] I like to spend time in a location to get a feel for the environment, and I don't want to affect it with my presence when I'm ready to record. I have very small packages of equipment and small microphones, which are more-or-less ignored by wildlife.

One of my biggest rules is to affect the environment as little as possible. I make a lot of conscious choices regarding where I record, thinking about topographical features.[41] Once I've placed the microphone rig, I press record and I leave it recording for perhaps one, two, or three days at a time, so that I'm not there affecting it.

Wildlife focused sound recordists will say you have to be there, and you have to see what you're recording, you have to put in the work, you can't just be somewhere and having tea, and talking to people while your equipment is doing the work. But, obviously, I think that's a bit limiting.

Usually, when I'm in a really remote part of the world, I will aim to record the environment only. Sometimes, people might occasionally come past and I will capture it, but this is not my main focus when I'm trying to record this sort of soundscape. I truly want to record nature as it is, as much as possible without it being affected. Wildlife takes several hours to get back to 'normal' after I've been to a place. You can start hearing the alarm calls dying down and then sometimes there will be some wildlife coming close and examining my equipment, and then just letting it be because they don't see any threat in it. Sometimes, they will interact – I have had baboons, bears, and elephants come very close to my equipment and sometimes break it and bite into my microphones, but that's how things happen in the field. You can't really blame them for that.

Occasionally, I will leave the rig in a liminal area, a place where people might come by to forage or to graze their cattle. I like the aspect of humans and I like the sound of community. I like the sounds of pockets of humans in an otherwise wild environment, for example, I recorded a small village in the Maasai Mara in Kenya, which was surrounded by the Savannah;[42] they could hear so much wildlife including hyenas and lions roaring at night.

The focus on conservation is often to parcel out nature and remove all of the humans, but hardly anyone considers the effect that this has on the humans. My Masai guide mentioned that in the 90s, they had to make a conscious choice to not shoot and kill lions anymore because the lion population was so depleted; it was really affecting the numbers of tourists visiting, and tourists were a significant source of income. Hunters were a large part of their community, but their livelihood was now gone. The government didn't offer any support for them to re-train, and they simply didn't know how to change their whole lifestyle, so unfortunately a lot of these hunters ended up being alcoholics and drug users, and it was a rough decade for them. So, this example, and it's not the only one I've encountered, has prompted me to think more about the human aspect in conservation and soundscape, even when I'm just looking for natural spaces.

Sharing Places and Spaces

I   *So, there is a lot of communication in your practice? It's not just about documenting what's there, it's about sharing that with other people?*

GV   Yes, absolutely. I think there is little point in making recordings and keeping them for myself. Likewise, it's not about recording absolutely everything, I always try to focus on a story or a narrative that I can relay afterwards, even if it's really open. I tend to focus on a few key elements.

If I'm out recording with a surround rig I'll quite often also record in stereo, because I can make a decision to focus on specific sounds. I think making deliberate choices and limiting my options is very helpful.

I   *I'm really interested in your approach to experiencing the environment, how you travel to a location, experience it for a couple of hours or a few days, and you get a sense for that space that you're in, and I suppose this is part of that decision process about enabling you to understand what you're going to record from that environment. Are you picking out salient features?*

GV   This is something I do without being aware of it, and I've done it all my life, so when I sometimes talk to people about it, they're surprised to hear that I stop and listen. I've always done that, so I don't control it, and I don't think about it too much.

I like to record spaces, and the acoustics of spaces fascinates me. I like to look for features in the topography of a place that influence how sound is perceived and how sound bounces around and is reflected. So, I like to look for gentle depressions in the ground where you can sort of get this parabola effect[43] and look for tree lines that will offer really drastic slapback delay.[44] I like to record at the edges of things, or where environments overlap because you get a bit more variation, a bit more variety there.

When I'm in a space, I usually have a local guide with me. I try to work with locals who have grown up there who can sort of present or show me how they feel about the place, and the most interesting aspects of the place, and usually it's a very interesting process from talking to people who have grown up in the one place and haven't really travelled much outside of it. Very slowly we kind of get to a point where we do understand each other, especially when you talk about sound. A lot of the people that I've spoken to were not aware that sound is of such importance. So, when I have had time to explain sound and listening to a local guide, the really interesting stage is where they begin to say "Oh, there's this new sound that I remember hearing when I was young, and I haven't really heard it in a while. Maybe we can go on a hunt for it". We get to these beautiful moments where we understand each other, not just conceptually talking about it, but being out and listening for something.

I can come up with my own anecdotes of growing up in a wild place and telling stories, so it's a really beautiful way to connect to someone. But also, it's a really good way to understand and to learn about an environment without having to live there.

I think it takes growing up in a specific place to really understand it. It's not just about experiencing it for a week or a month, or maybe even a year, you have to be a toddler, you have to be very close to the ground, and you have to be able to touch everything, and to walk and fall and get to the ground, and to climb trees, and to do all these things that children do to experience it fully. So, somewhere there's an answer to the question of what a place sounds like but it's not easy to convey. I prefer to ask people, "what does your chosen place sound like?"

The Feeling of Being in the Space

I   *I suppose it's a question of experiences? It's not just what it sounds like and what the vibrations are that are travelling through the air, but it's about how it feels?*

GV   Yes, I think so. I think sound has a really strong potential to recreate these moments and to convey these feelings. I think it's much better than video or photos, or even words. I think at times you have to speak to someone and to see their facial expressions and see their excitement when they talk about something to truly understand something. The next best thing is being there and recording these things, and then presenting them in a way that people can relate to.

I think the word 'to feel' describes it beautifully. It's not just about the particles of air moving around, it's the whole 'osmosis', this is why natural sound is so complex and so difficult to replicate using computers.

I   *So, you're not just recording sound objects, you're recording spaces of sounds?*

GV   Definitely. When I work on sound for video games, the most useful type of sound for me is a sound that's recorded separately from everything else, apart from its environment, as clean as possible. You don't want to have anything else present. But going out to a rainforest and only recording individual bird calls or wildlife calls is wasting a huge opportunity in my opinion.

Even if you bring back these as mono sounds and put them into your game engine, add your reverb, and pan them around, you won't get close to sharing even 1% of what the space sounded like. My main focus is still recording the sound of the space, and probably, the most exciting space I've ever recorded is a place called Langoué Baï in Gabon. It is a huge clearing in the rainforest, the size of many football pitches, surrounded by trees and in the middle is a mud pool where elephants and gorillas go to frolic in the mud and to dig for minerals.

It's very remote and was only first visited by Western explorers about 20 years ago. It took us about four days to get there, driving, taking a train at midnight, and walking for days in the rainforest.

We got there one afternoon, set up the microphones, and left them behind as a thunderstorm was moving in. I got these beautiful recordings of elephants screaming and trumpeting from the centre of the Baï, while this thunderstorm was approaching.[45] The thunder was reflecting from the trees at the edges of the rainforest, and the insects and rain, it was like a sonic painting. It's a really good composition. And to me that is the sound of a space. Everything brought together sounds just immensely more beautiful than just the individual sounds on their own.

Complex Networked Interactions of Sounds in Space

I   *So, it's all about the relationship between objects and the environment, and it's the relationship between all of the animals in the space and the insects, and the tree lines, it's that context that is what you're capturing in those recordings?*

GV   Yes, and the complexity of it. Because when you have these huge spaces, you get a myriad of reflections from the edges and from everywhere. You also get the effect of humidity on how sound is propagated, and then, you get all the natural relationships. How the animals react to the very powerful sound of thunder, to me this is perfection. And you can't ever expect to recreate that in a game engine.

I   *Do you think that there's a feel to it when you are positioning the microphone in terms of how the sound makes you feel? It's not conceptual, it's an emotional process?*

GV   I think so. To be honest, these days I don't really listen with my headphones anymore. I was fascinated with listening through my headphones to the environment, but I think I started to do that too much and I was out somewhere, and I was only listening properly with my headphones on through a pair of microphones. So, nowadays I don't do that anymore. When I set up equipment, I more or less have a feel for what it will sound like.

My favourite technique for recording is called 'tree ears', I like to place lavalier microphones on a tree trunk, and I like to think about it in terms of HRTF.[46] The tree obviously has grooves, and these features will impact or influence how sound reaches the microphones. So, I like to find some kind of an approximation of human ears on a tree. And I like to call this a tree's perspective.

I want to hear what that tree (which might have been there for hundreds of years or even millennia) has heard. It might sound a bit cheesy, but this is kind of my way of getting a feel for the environment. I want to approximate and capture the sound-scape from the perspective of a tree.

4.1.6 Steve Fanagan

Steve Fanagan (b.1977) is a Sound Designer, Supervising Sound Editor, and Re-Recording Mixer who works across a broad spectrum of projects in film and television. His recent credits include *Normal People* (2020), *Game of Thrones* (2011), and *Room* (2015). The TV series, *Normal People* is Steve's fifth collaboration with Academy Award-nominated director Lenny Abrahamson and was nominated for a BAFTA and won an IFTA for Best Sound.

He received an Emmy nomination and an MPSE Golden Reel Award for his work as a sound effects and Foley editor on the first season of *Game of Thrones*. In 2016, he was nominated for an HPA Award (Hollywood Professional Association), an MPSE Golden Reel Award, and International Music+Sound Award for his work on *Room*. Steve is a member of The Academy, BAFTA, the European Film Academy, Motion Picture Sound Editors' Guild, and the British Independent Film Awards.

Speaker Key

SF   Steve Fanagan
I   Interviewer

Building Sonic Spaces

I   *Composing spaces seem to be important in your work?*

SF   You're always trying to figure out what the world sounds like in the film. It always has a subjectivity, it's not direct – "See a thing, hear a thing" – it's rather – "What would this character be hearing and how would they be hearing it at this moment in time?" – and that's very much dictated by story and drama, performance and editing. It's really important to highlight that it's not just my work; it's the work of a sound team, in collaboration with our director and film editor. So, you're getting lots of clues and ideas from other people.

If we take the example of the film *Room*,[47] it's set in one single location for the first half. And that one location is very small and soundproofed. We began by trying to figure out what sounds might be in this space. There are visible story items – for example, at some point before the film is set the character of Ma tried to escape and smashed the lid of the toilet cistern over Old Nick, so we know there is a lidless cistern in there, which will create a specific sound.

We realised that it is essentially a prison cell, and Ma's been in there for at least seven years. So everything in there is going to be basic and old. That's actually great for us because it means that all of these objects are going to make very characterful noises.

Articulating Time Progression

One of the biggest challenges was to communicate the unfolding of time, which is also really important to the story. Morning is probably the best time of day for Ma and her son Jack because they know that Old Nick is not going to arrive until evening. So we designed it so that everything was a little bit calmer and more open in the mornings. I recorded in lots of ADR booths, mix stages, and studios to give a range of dry room tones and from these we chose a light, static, air, which we enhanced with EQ to boost the upper-mids and remove the low-end.

We wanted to communicate a feeling that, as the day progresses and Old Nick's arrival is imminent, the room is closing in on you. So, as the day progresses, the panning gets narrower, the tonality becomes richer (more low-end), and we begin to roll off some of the high frequency and upper-mids, which makes it feel claustrophobic.

We're also enhancing the sounds of the room objects, trying to make them a little bit more harsh and insistent, so that there's a sense of everything becoming less comfortable. It is about trying to do it in a way that's not obvious to the audience, that is felt rather than heard. It has to feel like a progression that's natural, true to the space, true to the drama, and taking its lead from the rhythm of the picture cut.

In terms of the characters themselves, there's a sense that Ma's life is very chaotic and that her son Jack is quite demanding. So, we also used Foley to increase the impression of them being on top of one another. When every little cloth move or hand gesture is felt and heard, there's a suggestion of heightened proximity (we did the same with breath).

Having a limited palette made us experiment and eke the most we could out of that small space. It becomes its own little universe.

Sonic Character and Performance

I   *It strikes me that you're creating characters out of all of the objects that are in the space?*

SF   Yes. You're always looking for the personality of something. When I go to the library and listen for sounds, I'm always seeking character. For example, if you think about an air conditioning unit, it's not the steady rhythm that's interesting, it's the moments of disruption where a rhythm changes or there's a shift in the cycle. Those little bumps are the most interesting because they feel individual to that object. When you figure out where to place such objects in a scene, their dramatic character can make a big contribution to the story.

When I first started to work in sound, I assisted a Foley artist, Caoimhe Doyle,[48] who was the Foley artist on *Room*. One of the things she aimed for was to capture the personality of the character's movements, how their weight shifts, and what that means.

She was able to perform both Jack's and Ma's feet in the same scene on the same surface and yet still have them sound entirely different for an audience! Understanding character is the thing that makes that difference.

I   *There is also a sense of liveness. Do you think that the fact you begin your career working with Foley artists has influenced the way in which you think about sound?*

SF   Absolutely! Caoimhe is world class at what she does. And I am very lucky that she was the first person in the film that I worked with, as she is someone who thought a lot about what they were doing and how they were contributing to a film in a very character-driven and aesthetic way.

It wasn't a case of thinking: "we need eight footsteps here". It was always, "what can we do with the rhythm and the cadence of this that says something about this moment in the movie?" That's really what you're trying to figure out every time you hit a new scene. "What can sound do at this moment?", building upon – what you read in the script, conversations with the Film Editor and Director, and what you've seen in the performance.

Storytelling Through Materiality – Timbre and Space

I've also worked in music production, where you make lots of decisions about how to mic up an instrument to convey its character – sometimes different characteristics for the contrast between the chorus and the verse. So, I'm also really interested in how sound works in a space. To my mind, perspective and the idea of acoustic energy is what it's all about!

Acoustic characteristics have a very clear emotional impact on me. We feel sounds; we don't think about them, we just instantaneously respond to acoustic properties, and they convey so much information about where the sound has come from – left or right, far away or close.

And *Room* is such an interesting study in that because suddenly you've got this limited setting and you're constantly on the search for those little acoustic cues that help communicate the feeling of the scenes.

I   *To what extent are you thinking about the audience when you're working?*

SF   You're naturally thinking about the audience because you're trying to give them a perspective. Filmmaking is all about choices. When someone chooses a lens or a shot size, they're pointing the audience in a direction, and with sound, you're doing the same thing. You're not thinking about individual audience members, but how sound is underscoring the performance and the drama that's happening on screen.

It can be a really simple, subtle thing. In *Normal People*,[49] when Connell and Marianne first kiss, they get close to each other, and the sounds of the world drop away. For the kiss, it is almost silent. As they got closer and closer to each other, we slowly shifted the mix – remove the birdsong, take out the ticking clock, add more breath and cloth movement – and it begins to convey an impression of proximity and the feeling that you're totally lost in this moment with these two people.

SF   I try to use my own experiences to guide my decision-making. If I'm getting lost in someone and we're about to kiss, I'm not really thinking about the birdsong outside. The world around has just got quite small and intimate; it's just about us. So that is what we wanted to emulate with sound. You're doing these unreal things but it feels natural.

Hopefully there is an honesty and truthfulness to that emulation, which recreates real human experience.

I   *So, you're thinking very much about proximity, scale, and the relationship of characters to the space itself?*

SF   No two moments in life really feel the same. If I give a guest lecture, I'm first conscious of the space – there's a little bit of shuffling from the class and there's maybe a hum from a projector – but once we get into what we're doing, your perception shifts and your awareness of those sounds goes away as you focus on your talk.

The idea that sound is static in a scene just doesn't make sense to me. It *has* to evolve because the experiential nature of cinema is as something that does evolve. As we move from a wide shot to a close up, there's a different feeling and sound needs to reflect that.

We're giving the audience an experience and we're pointing them in a direction, but if we're doing it right, it's in an invisible way. That's the power of cinema. If a film is well cut, you don't notice the cuts, but you feel the impact of them narratively and dramatically. If we can do the same with sound, then we're doing a service to the experience of watching the film.

Structure and Dynamics

I   *It also raises this question of this fine line between reality and artistic intention.*

SF   One of the biggest factors here is narrative, I like to think about it as an emotional arc. For example, in *Room*, the feeling of the day – opening it up or closing it down – helped us. There's a big moment in the film where Old Nick turns the power off. In the scenes leading up to that, the space needed to feel a little bit louder – everything buzzing and humming a little bit more – so that then when they wake up cold, the audience already has a cue, "this is much quieter than I'm used to". It's interesting how an absence of sound can communicate as well.

I   *Is articulating expectation and anticipation a key consideration?*

SF   When you're working on a scene, you tend to think just about what is useful to this particular scene. But, there is a point in the process where you have to listen to how that scene feels in the flow of the entire film. Sometimes, your instinct on the micro (scene) level might not be helpful to the macro (film level) when you reflect on the overall flow.

Being aware of the arc of the narrative is so important in this regard. You're trying to score with sound effects, Foley choices, and production sound in a way that feels like it's an overall arrangement that best serves the story.

For example, dynamics are just so viscerally useful to storytelling – loud/quiet is the most basic contrast you can have – but when you're trying to figure out the loudest and quietest moments, you have to remember that it's all relative. Your loudest moment doesn't necessarily need to be 120 dB, it could actually be something relatively quiet, but which seems loud compared to what precedes it.

There's no point in making every scene sound complicated just for the sake of it. If everything is complex, then you're just stuck at the same level all the time. Having something chaotic, followed by something suitably simpler, allows the audience to recalibrate and you get to prepare them for the next phase of the narrative.

Perspective and Acoustic Space

I   *Do you consider proximity when you're working? Ambience versus close-up sounds?*

SF   It was great for me to learn watching Caoimhe and her Foley mixer – Jean McGrath – make microphone choices. They're not just choosing a microphone, they're also making a decision on how to balance two or three microphones at once, e.g., a close mic, a room mic, and potentially a sub mic.[50] Jean is mixing these three channels onto a mono track, shifting perspective and making a real-time choice based on what's happening in the picture. This is the dual performative nature of Foley, it's not only walking the footsteps but also the performance of the mix between the channels. So, I'm conditioned to think in those terms when I go out of the studio to record, always considering how different microphones will capture different acoustic spaces.

This also impacts on the dialogue in a film. There is no getting away from the fact that one of our main goals in film sound is to present the performance of the actor. When a director and a film editor make a choice on a shot or take, it's more about performance and feeling, not necessarily about audio fidelity.

The ideal, for me, is to have a good boom as this allows two things: (1) it captures the human voice in a directional way, and (2) it also picks up something of the space that it's in. Yes, a shotgun mic is reducing the acoustic reflections to make sure that it's capturing as much of the human voice as possible, but it is also giving us some information about the space, and it's that acoustic energy which I believe is so important to this impression of reality.

I *It makes perfect sense that you're interested in using the boom, rather than the Lav, because you're interested in sound within an acoustic space.*[51]

SF Obviously, there are exceptions. If they're filming in an acoustically lively space Lavs can become very important components, and we're often using them to fill in gaps or help us with presence.

I *Do you feel that you're channelling different experiences of listening through the process?*

SF Definitely, this is why I keep talking about a subjective position, We're not recreating an objective sonic reality, but fashioning a subjective soundscape, which is much more to do with how we want to convey an impression of the character's feelings.

I'm always thinking about what we might be hearing that's off-screen, and there's something very important in how close or far away a sound is, in terms of how you then orientate the audience to the sound and the space. It's definitely always a 'point of view' and an experience that you're trying to give.

I don't necessarily know at the start how I am going to balance all of the soundscape elements to create that mood for the scene. It's a process of discovery. I'll often start by layering in the elements that you might expect to hear in an objective sense – e.g., "this apartment is right by a main road and it's four in the afternoon. So, we need the throb of traffic and a big heavy bus pass by", etc. But then I'll reflect on the goals of the story and think, "does any of that really matter?" This helps to guide our edit.

Process or Premeditation

I *I'm interested in the proportion of your work that is pre-planned and the extent to which it's more about the unfolding process of discovery? That is to say – finding out what should happen as you progress along on the journey of making – taking decisions that are contingent on other decisions.*

SF I've been thinking about this a lot lately, the work that you've done in the past always informs the work you're going to do and, as my first pass, I'll often do what occurs to me naturally. But as you delve deeper into the work, you're trying to bring out the version of the scene that is unique and true to this story.

You need to be open to the process and get to the point where the film itself is telling you what it needs. Ultimately, no two projects ever end up sounding the same. I've worked on films that use the same locations but have a totally different sound and feel because the film is asking for something different. As you get into the process, the film begins to communicate in some way, either through the people you're working with or through your own intimacy with it.

You realise there are possibilities that you didn't see the first-time round. For example, when I come back to scene seven after working on scene hundred, I might realise that there is a relationship. If I do something early on, it will pay off later, I can make decisions based upon what I know is going to happen in 20 minutes or in an hour.

I find that openness to process and the journey of making really inspiring; it's exciting too! You don't start the job with a preconception that rules how you should work. You start the job with an open mind and you do all that you think you should, and then evaluate. You do that yourself and with the people around you. And hopefully, through an iterative process, you figure out a version which is truest to the essence of the story.

Collaboration and Teamwork

I   *The collaborative element seems important to you. You seem to relish the dialogue and interaction with other creatives, whether they're working in sound as well, or a picture editor, or another role. Is it about the collaborative exchange of ideas?*

Filmmaking is the ultimate expression of collaboration, because you're a tiny cog in a complex clockwork organism. As long as you do what you think is honest for the film and communicate with the people you should, then hopefully you're doing something that's appropriate. And I really relish that.

No one can make a film on their own. You need a team of people, each specialised in a particular aspect of filmmaking. It's an example of the very best of the human community. There's something about filmmaking that's fascinating to me. It's wonderful to watch the credit roll on a film and see hundreds, if not thousands, of names who have all contributed to making this one film.

There are so many people who work on the same films as I do whose path I never cross, because we're coming in at the end of the process. Yet, somehow, we do collaborate with each other. We have an intimate relationship that perhaps neither is consciously aware of, but you do still inform each other's work. Tiny details of the art department and production design, entirely influence choices I'm making. And they may never know that, or they may be totally aware of it, I don't know, because we never have those conversations.

I   *People sometimes think that art is the product of one great creative mind 'the genius'. I like this idea you paint of the director being more of a curator, channelling everyone else's creativity that feeds into the filmmaking process.*

SF   The best directors that I've worked with allow you ownership of what you're doing, but help you refine it towards their vision. It is really valuable to bring the director in to give feedback on your work after you have disappeared into a scene for a few days.

Firstly, they haven't been in the weeds with you, so they have no idea whether something took 20 hours or ten minutes, and they can be blunt about it. Secondly, they've lived this project for years of their life already – through script options, funding, pre-production, etc. – and this can shed a different light on the project.

The director has a vision for their film, and they're open to the ideas that will enhance and help to underscore that idea. When I'm spotting with a director, I don't want them to say, "We need a door here and I want footsteps here". I want them to say, "This scene is about X, and this is how the viewer should feel. What can we do with sound that will contribute to that?"

Interpreting Notes and Critique

I   *Sometimes electroacoustic composers may see it as a negative thing to have an outside influence critiquing and guiding their work, they may see this outside influence as watering down their vision. I find that a really interesting discrepancy between experimental music and film.*

SF   Everyone gets notes on their work, even the Director! Lenny's [Abrahamson] attitude is that there is no bad note. But you do have to realise that the note may not actually be about the specific moment commented on. "I don't think that scene works" or "I don't like that beat" – might be more to do with the lead up to that moment. Something in the edit has perhaps stopped the flow so that it no longer makes sense.

Your reaction to notes cannot be emotional, it has to be professional, and it has to be open. Maybe ten years ago I would have been upset by them. But now I realise that it's not personal, I just need to figure out *why* you're giving me that note about this moment.

If you choose to engage other people in your work, whatever they tell you about it has meaning, purpose and significance. You can decide "I don't care, I love it". Or you can think, "Maybe I need to interrogate this more". My feeling is that the latter is healthier for your own creativity.

4.1.7 Jesse Dodd

Jesse Dodd has worked in professional film and TV as an Audio Mixer for over 25 years, with a focus on mixing ADR and Foley. She has worked for some of the most prestigious studios MGM, Skywalker Sound, Warner Brothers, Technicolour, and now NBC Universal and is one of the only Female African American Post-Production Sound Department ADR/ Foley Re-Recording Mixers in the industry.[52]

Jesse has over 250 credits and is a member of the Television Academy, CAS, MPSE, IATSE, and HPA. Selected credits include *The Bodyguard* (1992), *Blood Diamond* (2006), *Jurassic World* (2015), *Law and Order SVU* (2010–2018), and *Bel Air* (2022).

Speaker Key

| *I* | Interviewer |
| *JD* | Jesse Dodd |

Passion and Creativity

I   *What is the most fun, creative aspect of what you do?*

JD It's making it happen every day, under different circumstances that are the same. It's the same list of ingredients as an ADR and Foley mixer, but it's always a different outcome. For me, that is the creative genius of what we all do.

I   *It strikes me that in your role you're playing with this delicate line of reality versus fantasy. Do you feel you have freedom to explore and play with that?*

JD Well, let's take the two mediums separately. With ADR, I have a set of numbers, "air quotes", a room, an actor, and a performance. I have a producer and a director who wants this performance to be X, Y, and Z, as it was given on set. So, my job is to recreate what happened that day sonically. Here the creativity comes with the choice of microphone, where I am and we are at, and what am I working with to make this sound like what you see, as well as what happened on that day.

Foley is a whole different ball of wax because I'm working with amazing artists who come up with sounds to make the brain think it hears what it sees. It often gets more creative on the Foley stage because I watch in amazement as they make my job very easy.

The same is true for someone who loops well. I have actors who come in and "bing, bing, bing" and you go, "oh, my gosh!". Though others need a little more help, but that's great too because I get to assist in that process and make it happen.

It's a really beautiful meshing of two worlds – the creative side and the technical side. It's all about teamwork, and when I say team, I also mean the technique – the microphones, Pro Tools – everything that's involved must be in synergy to make it happen.

Collaboration and Relationships

I   *It seems that helping to guide other people's performance or encouraging them to perform is a core part of your work?*

JD With ADR, absolutely! We're all unique, and so you have actors who are really comfortable with ADR and those who really are not keen. And so that's when personalities become a huge issue. Not everyone can be an ADR mixer because you often have to get involved, and there is a hierarchy that happens on the stage, and you have to know when to speak up and when not to. There are ways to navigate those situations, and it becomes all about personality and interpersonal relationships.

I have to find a politically correct and compassionate way to say, "I need to do this again" and "Can you give me X, Y, and Z?". Because I also have a producer there and a sound supervisor too that's saying, "It's okay", and I'm saying "No, it's actually not okay. I'll give you that take, but let me get one more, just to back us up, to make sure that we're good. And give me X, Y and Z please".

It takes a special talent (the same is true for a dubbing mixer), because we are dealing with the nuts and the bolts of it – the actors, producers, directors, and sound supervisors directly. It's not a phone call, it's a relationship that we create. You want to make everybody as comfortable as possible *and* capture the best performance. I don't get the luxury of missing that take. I have to make sure that I'm on it technically as well, and you can't just throw a microphone up and let her rip, that's not how it goes.

Interpersonal relationship is everything. You have all of these pieces and they're all equally important.

Performance and Flow

I   *Do you feel that you are a performer when you're working with all of these people, collaborating with them?*

JD   I never really think of it that way because it is so innate to me. But I do understand the description though. I could be considered as one of the actors in the environment.

Coming up through school I was always technically inclined, and though I did plays, performance, and newscasting, I found that it was the nuts and the bolts, it was the plug, it was the key in the lock, and it was the motor engine that I could take apart and put back together. That was my thing.

There are some days that I go, "And what are we doing today?" And then the next thing I know, the day is over, and I go, "Oh, what just happened?" It's like I transformed into the scene and was another person - the ADR mixer or the Foley mixer.

I   *I love this description of being totally immersed in what you're doing and just being in a flow of work, so you're not necessarily really thinking; you're more just doing and feeling?*

JD   Absolutely. It's a synergy; we are all connected, people and tools. When I greet my board, I'm like, "Okay, guys, look here. This is what's happening today. How are you guys doing? Let's get at it! This is what we were born to do today, and you've got one job to complete. So let's do it!" And then I let go and I strive to create that vibe. Others have told me about this when they walk into my bay, and this often tends to be because we're all working collectively.

One time, a young man, part of the lighting crew asked me, "What do you think is the most important part of filmmaking or television making?" And I said to him:

There is no *most important* part. If you don't do your job, I can't do mine correctly. If I don't do mine, you can't do yours correctly. It's all equally important, so keep that in mind. Make sure you go all in, all of the time. Poop trickles downhill, man.

We're all a team. If I don't do my job well, this affects the rest of the sound team, you're only as strong as your weakest link. It's a group effort, and I have so much respect for everyone.

I have been honoured to have worked on Dick Wolf's[53] work, e.g., the *Law and Order's*[54] and the *Chicago's*.[55] Dick has been making television for many years and has put thousands of children through school single handed. Through what he's created and for those who have worked on his projects, it's had a significant generational impact in terms of boots on the ground.

It has been an interesting journey. To this day, there's no one who looks like me, doing what I do. Because I've looked for me. It's beautiful that there are more females, but I've had to navigate from a different part of my brain in order to 'succeed' (whatever that means to whomever). If I were to look at the industry hierarchy as nuts and bolts I'd be an absolute mess. Though my experience has been enjoyable with my sandbox from day one and that has made all the difference.

Social and Technical Demands

I   *What role does creativity and the potential of play have in the process and making? Is the creative at the centre of your work rather than the technical?*

JD   No matter what vibe I create or what my mindset is, if I can't use those faders and microphones how I need to, it doesn't matter. You cannot get into this on personality alone. It is not enough. And it just so happens that I can do both.

I often see younger people after studying, they're Pro Tools ready but totally missing the other side; you cannot be just technically savvy to do what I do. You might get away with it a bit more as a Foley mixer, because you're locked in a dark room with two other artists, but there's not an ADR mixer without the personality to go with it.

I   *One of the things that I really like is your description of how you get really hands on with things, really tactile. When you talk about plugging in cables, it paints a visceral picture. Is the 'hands on' physical aspect important for you?*

JD   Absolutely. I was blessed enough to start my career when there were still Magna-Tech machines,[56] with forty machines in a room! When they were about to start a reel, it was all hands-on deck, everybody grabbed two machines – three, two, one, push – and the mix stage started to go. I am so grateful for that. I didn't get to see it for long, but I learned about its flow, out of one bucket into the next, and even though we're now working digitally, the theory is still the same.

I think this tactility is imperative. I could sit back and just say, "make it happen, you guys, all I need to know is where my faders are", but that's not how I operate. I want to know what DANTE[57] is doing and how it's doing it, e.g., how many stages on the lot is it controlling, if my faders are sticking where on the board to go to reset them, etc. And not have to call my engineer every single time something goes haywire. So, tactility is everything.

I   *I think it also speaks to this idea of you being part of the system. How important is the process of recording to your workflow?*

JD   The recording process is just as important as having a picture. It's 100% or zero. I have people depending on me and the sound that I give them. I have to respect those who have cued it, those who are coming in and working to schedule to make this process happen, myself, the editors who are going to receive my stuff, the dubbing mixers, etc. Then, we have all of the finals, so the recording process, it's 100% or nothing for me.

Microphone Choices and Timbre

I   *I'm interested in digging a bit more into your intuitive approach. How do you make decisions about microphone placements and where to place microphones? Are you thinking analytically, or is it more about feeling and intuition?*

JD   Doing something for an extended length of time often becomes innate but if I think about when I initially started to do ADR and Foley, it was perhaps innate from the start. I never got to a place where I was thinking analytically, e.g., measuring the distance of microphones from a source. The closest I might have got to this is the choice of microphone, for example, thinking if I require a directional or condenser microphone.

It's about considering what I'm looking for and how the actors are performing. It's also dependent on the kind of voice over, for instance, an actor's voice may sound amazing on a Neumann TLM 103[58] for something outdoors, and then, the next actor in the same scene is just not registering as well on that setup, as the timbre of the voice can make a big difference. Over time, you learn how the timbre of a voice or sound will affect each microphone.

Sometimes, it is an investigation, and other times, I'll consider what they have used on set. So maybe I'll use a Sennheiser MKH 416[59] or a Neumann KMR 81[60] and see what it sounds like. And there are also times when I can't get the exact sound that I need, and I will talk to the sound supervisor and say, "this is as close as we're going to get in here right now". And there can be a myriad of reasons as to why it's not happening. I'm so thankful to editors who say, "we can deal with that. We're right in the pocket, I can take that down or I can raise it some". So, we all work together. It's become innate, it really has.

Creative Decision-Making

I   *Considering the process of decision-making, how do you know when it's right? When do you know that you can move on to the next take? Are you thinking analytically about the sound or is it the feel of it?*

JD   It's both. For ADR I must know that sonically what I'm hearing is as close as possible to what I can see. For instance, in a bedroom that has a lot of furniture in it, I must convincingly make my brain believe it's coming from a closet. Another is sync – if it's too long you can't time compress or expand it because it changes the timbre of the voice. (You can maybe do a little bit, but not too much without it changing). And also, match – I have to ensure that the lines flow. If you have a line in front of it and a line in back of it, and the audience doesn't know that's what has happened, then I have done my job correctly.

Again, everything is as equally important. It may depend on how the Editor or the Sound Supervisor works as to what's most important and what isn't on a particular project. Though, there is nothing more important than the other, other than capturing the sound in the best possible way.

Supporting the Actor – Physicality of the Space

I   *When you have actors in the studio, do you often layer in temp atmos*[61] *when they're listening to a line, to evoke the feeling of being in the space on location?*

JD   Absolutely! We have the picture up and audio running, and we play the scene as many times as required. And then it's whether the actor wants the production sound in their ear or not. Every actor is extremely different. Some will say "Let me hear the line and let me repeat it", even if it's a sync line, and sometimes, it works really well, and it'll fit perfectly. Others don't like the tip in their head because it throws them off, or they want the tip for timing or to keep an idea of where they are in terms of level and emotion. And if they don't want it, I just cut it up and drop it, mute it out. So, it varies; we let them watch the scene if they need to get back into the emotion of it. Whatever is needed, we make it happen.

I   *I always feel that one of the big challenges is when they're in a different place. If it's sonically different, they're hearing the scene, and then it cuts to a dry room, is that going to change how they react?*

JD   Interestingly enough, it is difficult to get actors to scream and yell. When you're inside you naturally want to use your indoor voice, and even when you want to yell out, it's still your indoor yell out. So, I have to say, "Step back and let it rip. Act like you're in a football field and let it rip!". That's often a big challenge when being inside. It changes your senses, and the body follows. It doesn't come out how it should, as the brain tells the body "I'm inside" and it tells the voice I'm inside.

I   *So, then it almost becomes physiological? You must help the actors to understand what their whole body's doing?*

JD   Aside from having them look at what's going on, we have them running, jumping jacks, etc., to get the breath required. For instance, in *Chicago PD*,[62] they're often chasing people, so we get actors jogging to be out of breath, huffing and puffing. If they're sitting down on screen, we will ask actors to sit, as it will pass through the diaphragm differently to if they're standing up and talking. It helps mentally a great deal too. It affects the throat, the diaphragm – the whole-body positioning is important.

We can also affect the voice in different ways, e.g., colds, allergies, tiredness, seasons, and times of day constantly change the voice. And you must know or at least be able to understand the effect of this when recording. The voice becomes a musical instrument; part of the scene, separate from the actor.

Sharing the Connection

I   *Considering flow, energy, and physiology of sound – what is it about the medium that draws you in? Why is it that you have chosen to work with this medium? And how has this informed your process?*

JD   I didn't choose it. It chose me.

Sound is a vibration. A type of vibration that you allow into your spirit, and your being affects your movement, consciousness – it affects everything. Music, which is a sound, is the only thing that I know that soothes my soul to the core. I have been blessed enough to allow these voices to become music to my soul. If I can take the good and peace that happens because of that, then I can give that back. You cannot give what you do not have.

So, it has become a whole healing process that continues to heal for me, through music, the voice, through the 'Os' and 'Z's'. And I couldn't be more grateful. I would do what I do for free and often have. The beauty is that I don't have to, but it is a life force for me.

I   *I think it's the reason why a lot of us do what we do, because we love it so much. And there's a reason why it resonates with us, whether it's in similar or different ways.*

JD   I agree. What I have allows me to give back in other ways, because it keeps me centred and grounded. I'm always looking at it like, "Wow, I get to come play? [laughter] Look who they let back on the lot! [laughter]. I have a pass?! Who knew!?" It's a gift. And I try to give that gift back.

4.1.8 Onnalee Blank

Onnalee Blank is a Re-Recording Mixer and Sound Supervisor. Originally trained as a ballet dancer and musician, Onnalee worked professionally in dance before an injury precipitated a change of career, first into music production working with Rick Ruben and subsequently into sound for TV and movies via work with Danny Elfman.

She has received ten Emmy nominations for her work on projects including *The Underground Railroad* and *Houdini*, and she has won four Emmy awards for her work on *Game of Thrones* (2012, 2015, 2016, 2019). Her recent film credits include *Thor: Love and Thunder* (2022), *Moonlight* (2016), *Apollo 18* (2011) and *Night Hunter* (2018).

Speaker Key

I	Interviewer
OB	Onnalee Blank

Creative Innovation

I   *What do you think are the biggest challenges in your role? As either Supervising Sound Editor or Sound Designer?*

OB   I always want to move on to the next job with a fresh creative headspace and not use the same tricks that I was using on the last show. That sometimes takes a little bit of time to adjust.

I was embedded in *Underground Railroad*[63] for almost two years, and it's certainly a challenge to move on to doing something totally different after that! But I'm always trying to keep it innovative. Pressing the envelope in different ways to create compelling soundscapes.

I   *What's driving this push for innovation?*

OB   For myself at the beginning. But as the project goes on, it's for the audience, because the goal is to create something that sounds different than anything they've heard before.

I   *So, you have a drive to create this fresh sound world?*

OB   Exactly. And to try to think of sound as a character in the film. Not just to put sound in the movie but to create a motif with sound design that works in the same key as the music score.

If you work with a Composer that's cool and collaborative, it really helps the process for sure.

Practices across Music and Film Sound

I   *I'm interested in that collaboration. You come from a musical background, and you worked in music production, do you find that there are similar creative approaches and processes that you can apply to the work that you do in film? Or are they completely incompatible?*

OB   I don't think they're incompatible at all. They are different as far as scope goes, you're thinking in stereo when you're tracking a musical record versus more discrete channels, 7.1 or Atmos[64] when you're cutting for film.

Nicholas Britell[65] is a composer I've worked with for quite a few years, as part of Barry Jenkins'[66] projects. That's a really wonderful collaboration. We have a weekly conversation about the work we are doing for certain scenes, we share stems of our work in progress, and we really try to write around each other. Sometimes, I'll say to him, "I need a trumpet in this key that goes with the drums I've created for the sound design in this moment". So, it becomes a mind-meld of ideas.

It's not always like that. Normally, it's a battle on the mix stage as sound design and music come together for the first time. As a Re-Recording Mixer, it's your job to look at it from a standpoint of what is best for the story of the project. Not to get caught up in thinking, "Oh, those are really cool sounds", and "That's a really cool piece of music, let's keep both and just play it all at 11. Screw the audience!" [laughter].

Balancing Detail and Context

I   *It seems like there's this incredible balancing act between the big picture and the smallest minute details?*

OB   Oh yeah, for sure. It can be very tedious work sometimes and every film is so different. It really depends on the schedule, the size and how many visual effects shots are coming in. It's also the job of the Re-Recording Mixer to keep an eye on the clock, and we have to get to a certain point by a certain time of day, without calling a producer telling them that we need more overtime.

So, pre-dub is a good time for each 'food group' in your film to have the time it needs to really focus on all the minutiae and dialogue and ADR, get that really singing properly, get it in a good space, and get your reverbs dialled in.

That's a good time to watch the film many times and think about the different ways you might be able to approach each scene. It's a nice way to wrap your head around the possibilities as the music and the sound design are coming together, "Maybe we don't have music here? Maybe we try to present the score in a less obviously musical way? How can we make the music more sound-like?".

It's really about knowing who your audience is and what type of show you are working on. We're not hired just to be mixing monkeys, "Raise that, lower that. Yes, sir". Our role is to think, "What can we give to the Director that they haven't thought of?"

It's really the last opportunity to be creative before it goes to streaming or the theatres.

Collaboration and Creative Distance

I   *It's a form of meta-creativity, in a way. You're taking creative elements that other people have moulded and shaped, and you're working with those blocks and shaping those. The challenge seems to be in this process of taking ownership of other people's creativity. It's at the heart of the collaborative filmmaking process?*

OB   Yes, it is. I think that collaboration is a really vital part of the process. I don't mix what I create. If I'm sound supervising and I've made some of the sound design, I actually prefer to have an Effects Mixer come in and mix the things that I've prepared, because I like to get a degree of separation from the material that I've been working on.

The mixer could just decide to take elements out and it's a good way to not be precious about your own material and to be open minded about other ideas that people might have. It's a really humbling experience to not mix your own material. Because there's no ego involved.

"Who cares, if someone doesn't like some of the sound design that I created?" That's not the end goal, the end goal is to end up with what is best for the movie and what is best for the story. How is the soundscape telling a story that's helping the dialogue?

We really are in a supporting role in the narrative. And it's a good thing to set step back. Even if sometimes it stings! [laughter]. You can end up in a situation where you think, "Oh, man I worked on that thing for two months and you're just going to take it out? Alright then [laughter]".

I   *At what point in the process do you think about the audience?*

OB   I go back and forth during the process. When I'm creating sound design and starting the project I'm thinking about the headspace of the character and how the sound is helping the characters in the scene.

On the mix stage, it switches. That's when I'm really taking more of an audience role, because you're sitting in the chair, almost as an audience member, making it for them at that point.

Emotional Communication

I   *Do you feel like you're thinking about the emotional content and the emotional expression when you're working? Or is that not something that you're worrying about too much?*

OB   I'm always thinking about that 100% of the time. Maybe that's just me, but anytime I'm working on a scene I'm always thinking how does this make me feel? What am I supposed to feel? What am I supposed to let the audience feel? Am I trying to play too much of it? Should we just let the dialogue be the dialogue? Or can we change the panning of that dialogue to bring the audience member closer to what they're saying? How can we play with levels? How can we play with different reverbs to create any sort of reality that you're trying to make in a scene?

I   *Is there a specific characteristic you look for in sound that predisposes it to that emotional communication?*

OB   I really like it if the actor is looking directly into the camera, which is obviously a clear aesthetic choice that the director made, to bring us close to that character. So I will think, "What can we do to portray the emotion that she's getting from the actor?"

You could:

- make it super close and personal,
- put the dialogue in all the channels to make it a voiceover moment,
- drop out all the other sounds and just have the voice.

Each decision could make the audience feel many different things. You just have to discover which is the right one! The only way to find out is to try them all and see which one works for the story.

Shifting Perspectives

I   *Are you talking about proximity and shifting of perspective?*

OB   Definitely. I love shifting of perspectives, I try to ensure the audience doesn't realise that I am doing this, and I don't want them to hear the process; ideally, they just notice that something's happening. It'll perhaps be how you EQ dialogue – add different low-end or high-end – or a different delay on the end of a word.

I'm a big perspective person. I like it a lot and when I was doing *Game of Thrones*,[67] I used to always get busted for doing too much perspective, "Onna, can you ease off the perspective"? [laughter] I'm like, "Oh man, come on?! That's the best bit!".

I   *Do you think that's your kind of calling card then? Is that your aesthetic? Your go to approach?*

OB   I love reverb. Reverb on dialogue especially, I think it helps fill up the room and does a lot to make dialogue super warm. And I think it helps convey that feeling of movie magic.

So, I love perspective and I love a nice reverb on dialogue.

I   *In a way you're creating that acoustic space in which then the audience can sit. Taking an individual point source and giving it an environment?*

OB   Yes. That's definitely true. Especially, if you're doing a period piece and they're in different halls and chambers with different floor types. Is it a stone floor? Is it wood floor? You can have a lot of fun with that as a mixer. You could spend hours choosing a reverb. Or at least I could! [laughter] "We've really got to move on? Oh, I'm still choosing some reverb here!".

I love using outboard gear, I still use it all the time. I get made fun of, but I like the option versus different plugins.

I'm also big into trying to create silence where silence is due and there's a lot to be said with subtleties and nuances of sound. It all just depends on the type of project that you're working on.

In *Game of Thrones*, we do a chair pass, so that anytime they move, we have to have the movement of the chair, every floor creak, and every subtle sound to have the environment really come alive in a unique way.

That's why a film like *Barton Fink*[68] is amazing, the walls breathe; when it's hot, the wallpaper peels off the walls and has a very unique sound; that movie is pretty wonderful to watch as far as the emotion of the soundscape and have a room come alive.

And I'm keen to take that kind of approach moving forward and to try to bring that into every movie. Obviously, certain comedies with Amy Schumer[69] are not going to really have the space for material like that. So, it all depends on the type of movie you're working on. And *you know* what type of movie you're working on, which is key.

Composing Spaces with Texture and Movement

I   *You've talked about shifting and moulding EQ but, I wonder, do you feel like texture is an important part of your practice?*

OB   In some ways. In *Underground Railroad*, there are so many bugs and cicadas, we recorded for weeks and days out in the Everglades, and even though it is a dense soundscape, you can't just use a stereo or 5.1 file to present your bugs because that's going to make it sound flat. So, a focal point of my practice is how to create a 3D sound with different mono files and panning and delays that spread your depth of field deeper, which I guess could be considered as texture. My feeling is that the field should be big and wide, and then you can really articulate the perspective that you want for the audience and the character.

I   *I wonder whether you're thinking about things more gesturally and about filling spaces with movement?*

OB　Yes, I just go with what flows and what sounds great. Things have to match from the dialogue side to the effect side. I really hate it when the Dialogue Mixer and the FX Mixer don't communicate on the mixing stage. You guys are a team, you should be using the same reverbs, you should be using the same EQs, you should be using similar mixing techniques, then the whole movie as a whole will be that much more cohesive.

I　*Do you think it's about composing space? You're creating and articulating spaces and environments as part of the process and that process helps inform other decisions that could maybe help inform or shift the edit in a way?*

OB　Yes, absolutely. When I work on Barry Jenkins' projects, I start designing when I read the script, and I send bits to the picture department when the editors are just doing their assembly cut.

　　What's really great about this is that it allows the picture editors, in their joy, to have new ideas. Because when she's hearing the sound come in, she might feel that she'd rather hold the shot for longer, while before she might have been going to tighten it. So, sound can really contribute to that process of learning to tell the story.

　　Communication from the beginning also means you can focus on the sound for a long time. You can try a million versions and have the time to get ideas flowing, because sometimes you don't think of approaches until towards the end of the project and you're like "Oh man, I wish I had thought of that earlier".

Space/Movement/Flow: Transitions

I　*Do you think that your interest in spaces (via reverb and perspective) has any relationship to your background in dance? In the sense of moving in space and being in space?*

OB　Yeah, for sure. I always think, "If I can dance to this and it flows, and it has grace and power", then it's working for the scene. Transitions from one scene to another should flow, whether it's a hard cut or soft, transitions are a big thing. It's the same in dance too. It's not necessarily about how high your leg is, or how much you're pointing your toe. It's really the transition from step to step that makes you a graceful dancer, and I apply that same principle with sound editing.

　　If every sound movement flows from one into another, it makes it feel like it was shot that way and the audience doesn't hear the edit. As long as it has beauty and grace and transitions are flowing, then I'm happy.

I　*I'm really interested in how your time as a dancer gives you a different perspective that other people don't have. There is a risk that we can think about sound as disembodied, invisible vibrations in the air. But a sense of embodiment and movement sometimes can be the most important thing to try and inject or convey through sound?*

OB　100%.

I　*Do you think about your work in terms of trajectory, energy, and movement?*

OB　Not so much when I'm editing, but I do when I'm mixing, that is all about movement and flow – just like in a dance piece – where's the climax? Is it at the beginning, or in the middle? It's the same.

　　I think of the soundscape as one unit, "Now it's passing on over to here and now it's going to this over here". It's almost like a river flowing through different forms on its course.

Expectation and Energy

I   *Within dancing, there's a lot of discussion of energy and levels of expectation (setting up expectations for future actions that are either followed through or subverted); is this what you are thinking about when you're talking about flow when mixing?*

OB   Different frequencies can make the audience feel activated and enthralled, while others can lower their energy, and this is what I'm thinking about when I approach a scene. You can do a lot of that work with EQ.

I did this a lot in *Game of Thrones*, because there are so many battles. How can you keep the energy going across a 45-minute battle scene and still keep the audience engaged the whole time? Watching these battle scenes many, many times before going on to the mix stage, we identified specific moments to reset, just like a dance piece "da, da, do, da". We reset the audience via sound, so as to have somewhere launch off too again.

The challenge is, how can you do that gracefully without the audience noticing and feeling "Oh, that was awkward".

As I mentioned earlier, before I go onto the mix stage, I watch the movie a lot of times and make notes on different ideas and concepts to try. These might be things that I think would be good for the emotion of the audience or for conveying the emotions of the character. Sometimes, I send those notes to the picture editor to suggest changes that will allow us the space to make these sonic moments happen.

Certain things like breathing are really powerful, you can do a lot to play with the subconscious mind using breath. Even if it's not registered on screen, one of the things that I like to do is to shift the panning. You can have the breath wide and make it part of the music, then bring it in and make it super mono, and sometimes, it makes the audience just shift in their subconscious minds.

Performance as key to Practice

I   *I think there's something really interesting about the synergies between your experience as a dancer and movement. Can I ask what it was about sound, and in particular working with sound, which brought you into this practice?*

OB   While I was a dancer, I also played music, and when I got injured in dance, I still loved music, but I didn't know if I wanted to play. So, I thought, why don't I learn how to record music? So, I worked in the music industry for a while before I got a job with Danny Elfman.[70] And he opened my eyes to film and its creative process.

I was drawn to mixing because it is performance based. People are constantly staring at you on the mix stage, you can't really make a mistake. So, there's a lot of pressure and I think that I always perform better under pressure than if no one's watching.

When you're mixing, you're always performing, although I didn't consciously realise this until a few years into working as a mixer. I realised it was like being on stage in a weird way.

When you want to try stuff on the mix stage, you sometimes have to turn around and say, "hey, I'm going to try this, if it doesn't work, that's fine, we can try something else". You just have to warn people that you might make a mistake.

I   When you talk about this idea of a dance and space, I wonder, if you think is it also about choreography? You're bringing together these different elements and so it's about crafting an ensemble piece from solo elements?

OB   It's definitely about an ensemble piece, sometimes there's too much ensemble [laughter], too many people dancing the same part and it's okay to thin it out and build it in a different way.

Sometimes, you might have to push back against the structure of the score as written and let the scene build in a different way. But also, a lot of Sound Designers and supervisors get stuck on their own ideas, "This is how it's been in the edit the whole time and we're not going to change it" and I really hate it when people say that because, well, "Why? Why won't we change it? Why don't we try something new? We're here at the mix to try stuff."

Creative/Bold Alternative Strategies

I   I suppose, because the context changes when you start to bring all of the elements together, you've shifted the balance of everything. So, you can't just have everything as it was individually.

OB   That's totally right. Yeah. You sometimes work with people who are less open to trying new material, and sometimes you just have to accept that and let it go, but I'm always pushing to get people to try new things, even if just one thing.

And it's important not to be afraid to be bold. Have really loud moments and soft moments, and bring the audience in. I just finished working on a film where I kept saying "Be bold or go home. We're here because the audience wants to hear this be badass. So let's play".

Sometimes, there's a fear to be different, and I think maybe I thrive on pushing against that. I'm just starting work on the new *Mufasa: The Lion King*,[71] but trying to think of it, not like any of the other *Lion King's*. I'm creating early sound inspired by the storyboards that is completely different than anything before. But, it is challenging because you have to remember that it is still a kid's movie, so you can't be too scary.

Another example is *The Underground Railroad* which we just finished working on. It's based on Colson Whitehead's novel and takes place in the southern American states. At first, I was making it sound like a traditional historical period piece, then a month into working, I realised, "I think this is actually a horror movie". I mean, slavery is a horror, it's terrible.

So, I decided, let's not have any birds. We deliberately didn't put one single bird in there at all. Yet you're in the south and clearly, clearly, there are birds there. Instead, we just used vultures and fox calls and raccoons wailing in the night, all weird animal sounds that fall in similar frequency ranges to birds. If you just watch it, you wouldn't really notice, but subconsciously you feel, "Something's off about this", and I thought it worked brilliantly.

4.1.9 *Peter Albrechtsen*

Peter Albrechtsen (b.1976) is a Sound Designer, Re-Recording Mixer, and Music Supervisor based in Copenhagen, Denmark. He works on both feature films and documentaries, and his credits include *Dunkirk, Darkland, True Conviction, Bill Nye: Science Guy*, and *Land of the Free*. Peter is a member of the Academy of Motion Picture Arts and Sciences and has collaborated with a variety of globally acclaimed musicians including Antony and the Johnsons, Efterklang and Jóhann Jóhannsson. Peter is also a member of the US Association of Sound Designers, MPSE.

Speaker Key

I   Interviewer
PA   Peter Albrechtsen

Sound as a Storytelling Tool

I   *Do you find that there are certain projects that you are attracted to working on more than others?*

PA   Sound is an incredibly important part of the film with massive creative potential for storytelling. So, I always want to be able to approach sound as something that is a key player in telling the story. I'm naturally attracted to projects, where directors really want to explore how you can work with sound as a very essential part of the storytelling. So, I'm very fortunate and privileged to work with a lot of filmmakers, where usually, I'm a part of the project from very early on from the script stage, so that I'm really a part of developing the whole film. I get to be very creative and have a big part of the whole storytelling, the whole drama, and the whole emotional content of the film is very much based on sound.

It's really important to get to work with people who have the same creative approach to how to work with the sonic language of the film.

I   *Do you think there are more directors and producers who are aware of the power of sound? And do you think people from the sound department will be brought in earlier on in the film?*

PA   There is a change towards more focus on sound. I think there's still a long way to go, but I feel that the new generations of filmmakers, directors, and producers are much more into sound and really know what sound can do for a film.

And that goes for both fiction films and documentaries. I've worked across both, but I find it really amazing how documentaries have developed this amazing film language where it's not just about words, but that the visuals and the sound of the film are incredibly important.

Usually, just 10% of the budget goes to sound, and the rest just goes to the picture. But I did this film called *The Killing of Two Lovers*[72] last year, and 50% of the budget went to the sound. When I told that to some people afterwards, they were blown away. It is often much cheaper to do things with sound, but still, there's this whole absurd lack of balance between what's spent on one thing and what is spent on the other.

I also feel there's more and more really interesting movies being made around the world. I'm a member of the Academy,[73] so I get to see films from all around the world in the international film category. Just watching movies from all around the world where there's so many filmmakers who are clearly, very ambitious, not just with visuals but also with the sound.

Experimentation and Emotion

I   *In terms of your own work, do you have a particular aesthetic that people would recognise?*

PA   I think people hire me because they want to experiment and play around with sound and explore what sound can do. I am also fortunate to get to work with a lot of directors several times, and in this case, you also develop a common language about sound. When you work together for the first time, you talk a lot, and you try and find out, how do we communicate about sound, and what common language and terminology can we use?

But after that first project, you get to know each other very, very well. So much so that you almost don't have to talk anymore because it's so unspoken, in a way. It's so much about emotions, when they're in the room with me, and I'm playing back something, and I pause after showing them a sequence, they hardly need to talk, because I know what they feel.

Because sound is subjective, if you have a room with ten people and you play back a sound, you're going to get ten different opinions of what that sound means. There's no objective truth about sound, it is totally subjective, which is quite amazing. That makes it hard to talk about sometimes. But it's also really wonderful because it means that you have to be very personal in your communication about sound.

If you really want to explore what sound can do, it quickly becomes a very intimate and emotional discussion. You really have to open up about how you're feeling and react to certain things, the visceral reaction of hearing a sound, and I find that really interesting.

Meaning Through Feeling

I   *Do you focus more on the feeling in sound, rather than any meaning? Or do you think that meaning comes through the feeling?*

PA   I think the meaning comes through the feeling, I love movies where they open up and I can be totally embraced by them. It's like inhabiting a different world and sound is incredibly important in this regard. It has this ability to create depth and dynamic, but also connects so emotionally with us.

The more precise and impactful the sound is, the more emotional the sound is, the more it connects the audience and the more the audience connects to the film.

It's also a way of sparking the audience's curiosity. Every sound I edit into a film has to tell a story. Why else would it be there? If you have that approach to sound, then it also heightens the audience's emotional attachment to the story, because the audience feels that everything is telling them something important.

I   *Do you want people to be aware of and pay attention to the sound because the sound is sending a message that they should be listening to?*

PA   I think people are much more aware of sound than we think they are, but in a subconscious way. They register everything but wouldn't be able to write down what happened after the fact! We don't generally have the cognitive language to be able to describe it in words.

It's very important for me that the film makes you listen and not just hear. There's so much noise in our everyday world that we constantly filter out, just to be able to survive psychologically. So it's important that every sound in a film is controlled and carefully designed to ensure that there are no superfluous elements that might trigger someone's brain to 'filter out the noise'. Otherwise, you distance them from the story, and the end result is that you push them away, rather than embrace them into the experience. If your sounds can embrace them, you really make people's emotional reactions to a story so much stronger.

I   *That's a really beautiful description. This place of the cinema is a space where you mediate, your brain doesn't have to work to block things out, and it's creating a generous experience for people in that sense.*

PA   Yes, it's also quite a lot of pressure on you! [laughter].

I   *It's interesting what you say about the subconscious nature of it that people are impacted but they don't necessarily recognise it.*

PA   Sometimes, we might wish that the audience is more aware of the work that we do. But actually the fact that the audience can't hear my interventions makes it possible for me as a Sound Designer to create an emotional experience in the cinema.

Horror films are so based on sound. And a lot of great directors of horror or suspense are brilliant at using sound. Hitchcock,[74] for example, wrote sound scripts for his movies so that every scene had a detailed plan of how it should sound. It's amazing, so precise, and so many of his films pushed the envelope and were really experimental sound wise.

Reality and Emotional Authenticity

I   *To what extent are you conscious of twisting and manipulating reality within the work that you do?*

PA   I prefer to talk in terms of emotional authenticity. Somebody has set the sounds to feel emotionally authentic, but not necessarily physically authentic. Sound has to tell the right emotional story. So, I spend a lot of time finding the right sounds. I make long, long lists of sounds I have to collect, and I spend a lot of time either recording those myself or getting recordings from around the world.

What I'm building and shaping is definitely my interpretation of this world, but a lot of that work is just done intuitively. Afterwards, I may realise, "okay, I did this because of that". But primarily, it is guided by feeling. It has to feel right, but it doesn't need to be objectively real. For example, when someone slams a door in a film, I'll often cut in a little bit of a hand grenade, because it makes the door slam more powerful. People don't hear the grenade but just think, "Oh, this person is really angry". It feels emotionally right. But it's definitely not a realistic sound.

I   *I love the example of doors, because they are so simple but can be so emotive, as well as having this very powerful communicative role and function of articulating movement between spaces.*

PA   I often go back and forth between using exterior sounds of a natural environment and an internal subjective soundscape from the characters' perspective. So I'm thinking, what are the sounds that are surrounding these characters, but also what are the sounds that would be inside of these characters? What are they hearing? What are they feeling? How do I communicate that with sound?

Process or Premeditation

I   *When you're working on a film, will you have a plan from the start? Or will you create different interpretations of a scene as you go along and flip back and forth between choices to decide if it works better this way or that way?*

PA   I often have a plan from the start, but maybe 20% of that plan actually works. Then, I'll get another 20% when I see the dailies[75] from the shoot, and another 20% coming in when they start editing picture. The last 40% emerges through the process of sound editing at the end. So, it builds over the project, but the important thing for me is being part of that process early and being able to tell a director, "I think we need to try and shoot the scene in a way where the characters are quiet. Because I think it'll really change the feeling of the whole thing." For example, you have a couple that's breaking up. And instead of them screaming at each other on a noisy street, then let's put them in a restaurant that's super quiet and they're whispering to each other, [whispers] "I need to leave now, this is terrible. I can't be in this relationship anymore". The whole feeling will be so different and the more you think about those things, from the script stage, that will inspire the actors to act in a way that is responsive to and considers sound as well.

I   *There's so much that can be done with the voice in terms of articulations and enabling actors to have the opportunity to integrate sonic ideas into their performance is a fantastic idea.*

PA   I do that more and more, and I think it's incredibly important to have some interaction with the actors. Some of my best experiences have been where I'm a part of the creative process around the script, and this means that it inspires the actor to think about sound and listen when they're acting. The Coen Brothers[76] always write in sounds, just small things – he's sitting down on a bed, it creaks, he's switching off the TV, and it has a click – just small things, but they make the actors aware that they have to listen and not just hit their marks or look in a specific way, but also listen.

Texture and Materiality

I   *As you said, we're always listening to sound, so we will naturally respond to things differently when they make a certain sound. To what extent do you think materiality is important to you in your creative practice?*

PA   I usually work with an assistant who helps me – recording sound effects and cutting some sound effects for me – and I often work with the same Foley artist and mixer. If you ask any of them about working with me, then they'll all say, "Peter talks a lot about texture".

I'm obsessed with the texture of sound. I find it so incredibly important. I feel a sound can be made more interesting if you record it in a room a little away from the source, because suddenly it gets a new texture from just the reverb in the room. I really feel that.

The feeling of the texture of the sound is something that I'm obsessed with. I love doing all kinds of small touches to a sound, so that it has really the right texture.

I hate anonymous sounds, as I said before, I feel sounds need to tell a story. If I'm making a horror movie, then I want the car pass to be a screeching car pass "vr-iiiiiiinnnnggg". And if it's a comedy, then it needs to be a weird car pass that has some funny story to tell, "hhuh-he-he".

If it's just flat, boring, anonymous texture, then it doesn't tell any story and I don't need it. Often, story comes from the texture.

I   *When you talk about texture, it has this strong emotional content, it's about the feeling of the texture, and do you think that the feeling comes through the texture?*

PA   Yeah, absolutely.

I   *We have texture, we get feeling, and through the feeling, we have the meaning.*

PA   It's a poem about Peter Albrechtsen's sound work [laughter]!

4.2 Electroacoustic Music Interviewees

4.2.1 *Hildegard Westerkamp*

Hildegard Westerkamp (b.1946) is a Composer, Radio Artist, and Sound Ecologist. After completing Music Studies at the University of British Columbia in the early 1970s, Hildegard joined the World Soundscape Project under the direction of R. Murray Schafer at Simon Fraser University. From there, Hildegard's career in soundscape composition and acoustic ecology emerged, undertaking a Master's in *Listening and Soundmaking – A Study of Music as Environment*, alongside teaching acoustic communications courses until 1990 at Simon Fraser University with Barry Truax. Hildegard was instrumental in founding the *World Forum for Acoustic Ecology* and was chief editor of the journal *Soundscape* from 2000 to 2012. Hildegard has written an abundance of articles and texts addressing soundscape, acoustic ecology and listening, alongside producing various soundscape compositions, soundwalks, soundscape workshops and lectures. She has composed film soundtracks and sound documents for radio and hosted radio programmes such as *Soundwalking* and *Musica Nova* on Vancouver Co-operative Radio.

Speaker Key

HW Hildegard Westerkamp
I   Interviewer

Inspiration from Sound Material – Practice Embedded in Lived Experience and Activism

I   *Do you think there's an underlying character that draws you into particular sounds or places? Is there something in common between the sound materials that inspire you?*

HW One thing in common is that all of my pieces somehow relate to what I'm doing, thinking, and/or has been happening – the lived experience. And so, this interest is strong enough that I want to speak about it and create works that help to communicate this.

For instance, while I was with the *World Soundscape Project*[77] at Simon Fraser University, I wanted to learn more about our analogue Sonic Research Studio and its equipment. While there, the idea arose to create a work with the sentence "when there is no sound, hearing is most alert" (Kirphal Singh) from Murray Schafer's book *The Tuning of the World*.[78] This grabbed me philosophically and in relation to sound activism – raising awareness about silence and quiet. From this developed the piece, *Whisper Study*[79] using a recording of the whispered sentence and using analogue techniques. While in the studio, undisturbed and without interruption, considering silence and what happens when we are in a quiet state, it was within this listening context I discovered that certain sounds heightened the imagination – some of the sounds that emerged from processing the whispered voice were very liquid like, almost like a river sound. They reminded me of sitting by a creek, hearing different sounds in water, almost like voices, hearing the voices of ancestors.

In addition, during the process of composing *Whisper Study*, using only three sources (which included a recording of an alphorn from The World Soundscape

Project collection) and applying the same techniques provided a framing and prevented me from getting too confused with the technological potential of the studio. Often, this constant framing is required, and it's why I've never been drawn to abstract composition and electronic sounds, as there's a danger of getting lost in space, and then, I would lose my compositional inspiration. There are certain techniques that I use quite consistently, as I am often overwhelmed by the multitude of processing tools available in the technological world. And so, I like to have boundaries in my approach.

With almost every piece, the topic, interest, and idea are the starting points, and I then begin to work with the sound, or I go out and record (or search for the sound required!). The context of environmental sounds and their language, what they have to say and bring with them into the studio is why I compose. It is the focus. It's the conversation between the sounds and my own compositional imagination – the sounds themselves guide me into a piece and its structure. The sounds are the words and I'm using those sounds to speak about something. The interest lies precisely in that interaction between the recordings, the environment at that moment, and the work in the studio when you're working with those sounds.

Sonic Communication – A Dialogue with the Sounds

I *In the sense that the sounds themselves can have agency, is this more of a dialogue between yourself and the material, rather than traditional compositional thinking, where the creator layers out the sounds?*

HW Yes, and it's based on the fact that I had worked with Murray Schafer on *The Tuning of the World* as a researcher, and with the World Soundscape Project. Our main focus was the quality of the soundscape and the ecological issues connected with it. When I discovered Schafer's work, I found something of vital importance to me, but of which I had not been aware up to that point. My activist self wanted to improve the world, and so, the work in the studio literally grew out of that. This led my life through research to focus on activism and acoustic ecology work, and later creating the *World Forum for Acoustic Ecology*.[80]

This compositional work is part of a much larger conversation regarding listening and understanding *how* we listen and relate to the environment through our ears. There is a message in every sound that we hear. For instance, the sound of a car passing tells us a huge amount – possibly the mood of the driver, the surface of the road, the quality of the tyres, acoustics in the street, etc.

Openness and Honesty

I *It's really interesting that you mention anti-abstraction, if you want to understand what someone's doing and how they're seeking to communicate with people, then you can't remove the works from the context in which they're created or the sounds.*

HW I could not work any other way. In the beginning, I would always talk about the process of the work in the studio, about mistakes I made or things I discovered. Though I noticed, especially in the musical community, that there was often either astonishment or judgement of being too personal. The work, *Moments of Laughter*,[81] a recording of my daughter's voice, from shortly after birth until about seven years of age, traced the development of the baby's voice to language, along with a soloist who provided extended vocal techniques imitating the child. I had noticed discomfort from some audience members. Some commented at the time that "it's pretty close to the skin, some of it, right?".

I think it has to do with comfort or discomfort with a kind of intimacy. The baby's voice is very truthful; it tells us exactly what is going on emotionally. There can be a deep psychological resistance towards people wanting to open up. We're not encouraged even when we're younger to enjoy voices that go beyond speech, voices that express difficult emotions. Indeed, it can be very hard to listen to and actually take in. And I think that piece for some listeners touches on this – it may be too expressive and uncomfortable for many.

And so, the way in which I work is that the life process leads me into a piece and the materials into the meanings and structures of the work. This is very close to how life's journey is perceived in India and I think that's why I love Indian culture so much. It goes something like this: "On a journey, it's not the destination that's important. It's the process of getting there that's important".

I   *Do you consider that this is an issue within the contemporary music community? It can still be very conservative about the way that it frames and values judgements around what is acceptable and the way in which works are often developed from a concept over the process of discovery?*

HW I think there is a very deep conditioning from several sides culturally, certainly in Europe, that you must produce a big work – the cult of the genius. I grew up with a sense of Mozart[82] being this incredible genius, and so to think that you're a composer seems impossible, when comparing to him. There is also a strong product orientation, where we must come up with something – we are getting a product, the recording, and are bringing this item into the studio. One could possibly call it a deep capitalist attitude, we have to do something with it, and then, we have to publicise it and sell it. This kind of framing does not offer much room for letting compositions emerge from within a creative process.

Composition as Communication

I   *When you're composing, are you thinking of the listener, and to what extent are you thinking of the listening context of the concert hall? Are you considering the listening audience at the end when you're in this process of discovery through the materials?*

HW I think it's both. I want to share what I'm hearing. And what I'm experiencing is so exciting that I want others to hear it. There's again a sense of conversation. *A Walk Through the City*,[83] commissioned by the CBC for the new music programme, *Two New Hours*,[84] was about the Downtown Eastside in Vancouver, an area plagued by poverty, addiction, and homelessness. At that time, I was involved with Vancouver Co-operative Radio, which was stationed in that region. And so daily I would go there, walk across this small plaza and witness the suffering. And I thought, "the contemporary music audience of the CBC needs to hear about this".

I wanted listeners to hear the soundscape of the Downtown Eastside and based the piece on a poem written in response to the area by my then-husband Norbert Ruebsaat, *A Walk Through the City*. I wanted to challenge the conventional listeners of the CBC as I felt that much of the audience had very specific expectations of what new music should be. And so, I wanted to rattle those expectations a bit and be public about it.

Listening as a Process of Discovery

I *I think for me, one of the things about your work that is so powerful is the ability that you have to guide people to listen in new ways and to share a perspective of listening that you have discovered or captured that carries them into that world. An ability to share sonic perspectives – do you think that's fair?*

HW Yes, it's fair. Listening is always new, and when you're in the process of listening to the environment, this is always a discovery, because nothing repeats itself. The environment always changes, and it's the way of listening which guides you through. When we participate in soundwalks, the discussions afterwards are always incredibly interesting because people have new discoveries repeatedly.

   This is what is so inspiring about connecting to the world through the ears because you have to be prepared for surprises and the unexpected. There are multiple social and environmental issues, and so we often question, how much are we willing to open up to the environment, and how much is the environment giving us the opportunity to be open to it? This is an ecological question.

   Our experience has been when we've gone on sound walks that you never forget the area that you listened to on this walk. Because you're spending a type of attention that you can't in daily life, when we're often so busy and preoccupied. If we are able to listen to the environment – as much as we do to whatever we are preoccupied with – we may also become aware that the characteristic of a soundscape in which we're living is always in conversation with us. We need to understand the complexity of this and that it takes much awareness to understand how we converse with our environment, or how it converses with us. When the world suddenly became quiet at the beginning of the pandemic, many people got an inkling of the acoustic violence our lifestyles routinely impose on the environment. When that suddenly stopped, the natural environment, even our own voices and footsteps, became audible in many places. The pandemic has highlighted something acoustically that now carries much relevance in dealing with the climate crisis: the importance of listening to the environment

I *Every time you listen to a piece, you're listening in a new way even if it's the same piece, and it's this shift from considering sounds or works as absolute abstract boxes that are outside of context, but actually seeing them as part of a larger scope and an unfolding process.*

HW Exactly! A piece offers a moment in time during which you are allowed to spend focused time with it – it's a framework to allow for intense listening. And sound walks are just like that as well.

Formats of Presentation – The Role of Storytelling

I *So, when you are listening and drawn into particular sounds of a place and/or environment, how does this inform whether the sounds form a composition or soundwalk? How does the environment inform how you choose to present it afterwards?*

HW Many of my other works are commissions, radio programmes or sound journals. Another example *Kits Beach Soundwalk*[85] came literally out of nowhere. It was not a commission, and it was not planned. At that time of my life, I was writing my thesis about background and foreground music, music as environment, considering quietness, technology, and my role as a woman in the music scene, and I captured a recording on Kitsilano Beach in Vancouver on a quiet evening. Coincidentally, I had been on a nearby island recently and had witnessed 'fingers' coming out of the barnacles feeding, and heard the sound that it produced. I didn't have a recorder at the time. And so, when I was on Kits Beach, I thought, "Oh, that sound, there it is! But here's the city, and it's loud, and there is a big hum around it". Using the recording, I began experimenting with this, while also experimenting at the time with high frequencies, and brought both together in the work. The work, both the soundmix and the narrative, was simply an expression of all I had been thinking about at that time. Initially, *Kits Beach Soundwalk* was a performance piece. In concerts, I would perform the spoken text live over a microphone, similar to a radio programme. Later, I recorded the text for the CD publication.

An example of a sound journal is the *From the India Sound Journal*.[86] This was my attempt to deal with and express my marvellously overwhelming experiences in India. It is not a composition. I presented it as a performative work for a while, in which I combined field recordings with live spoken narrative. And so, there was a period of storytelling in my compositional approach.

Similarly, *Cricket Voice*,[87] a piece in response to my experience of a desert in Mexico, was developed many years after capturing the recording of a solo cricket. The very close-up recording in the desert was a miracle occurrence, and I didn't know what to do with it. I had been sitting in the dark for several nights with cricket sounds surrounding me, and then suddenly, one cricket came forward like a soloist directly in front of the microphone. It was astonishing. And then of course, the question became, "What do I have to say about this cricket? What does it say? And what do I even know about it?".

And so, with this cricket sound, which was so special and beautiful, I felt that I wanted to honour this small animal and its role in the ecological system. When I altered the speed of the recording, a most remarkable throbbing sound emerged, almost like a heartbeat, and another one like a chorus of human voices. They were big sounds that – so I felt – had the power to amplify and honour this tiny creature and its importance in the natural chain. These in conjunction with the original cricket sound created – in my mind – a balance between letting the sound itself speak and revealing its inner deeper sound qualities.

Complexity of Listening

I   *Would you say that you're considering honesty and integrity of wanting to retain the character of the sounds that you have recorded and aren't masking their voice with your own? That you're joining with the sounds and helping them to speak.*

HW   Yeah, though I don't experience it as an honesty, I experience it as something that is necessary. It was something that requires balance between reality and imagination, often placing side by side the processed sounds and the original and comparing them. We're always interpreting what we're listening to. So, the processing of a sound to me is symbolic of what happens when we listen. The listening process is inherent in this balance, and it has ecological implications – we need to balance what we hear in the world with how we interpret it. The original work with the *World Soundscape Project* and the knowledge I had gained about sound, listening, sound behaviours, how our ear works, etc., informs all of this. As I grow older, the complexity of listening grows larger and seems never-ending. The sonic world continuously offers us inspiration if we are open to noticing and hearing that.

4.2.2 Annette Vande Gorne

Annette Vande Gorne (b.1946) is an Electroacoustic Composer based in Belgium. She studied with Guy Reibel and Pierre Schaeffer at the *Conservatoire national supérieur* in Paris and worked with François Bayle at the GRM. In 1982, she returned to Belgium and founded the *Métamorphoses d'Orphée* studio in Ohain. Since 1984, she has curated an annual series of concerts and the acousmatic festival *L'Espace du Son* in Brussels with the acousmonium, an 80-loudspeaker orchestra system.

In 1986, she became Professor of Acousmatic Composition at the Royal Conservatory of Liège and has since taught in Brussels and Mons. In 1995, she was awarded the *Prix SABAM Nouvelles formes d'expression musicale* (SABAM Prize for New Forms of Musical Expression), and in 2021, she received the *Thomas Seelig Prize*.

Speaker Key

AVG Annette Vande Gorne
I   Interviewer

Sound and Inspiration

I   *How do you know when a sound will inspire a new piece?*
AVG I always start with a concept, and there is always a relationship between that concept and the sounds I choose. Pierre Henry[88] told us, "I classify sounds and then I compose." And what he means by this is to not think about sound in an abstract way, but to classify the sounds according to either their energy types or their poetic resonance. So, we can think both about the sound itself and the meaning of that sound.

I am guided by the meaning of the sounds that I hear, in relation to the concept of the project. For example, two days ago I finished a new piece that contains a requiem section. Responding to the tropes and the traditions of the requiem, I took an old vocal piece and layered up the voice to create a slow-moving choral assembly. I knew the subject and the title before I made the piece, so all of my decisions were informed by this idea! Often, I find that the title exists, but I don't know anything about the piece yet.… [laughter].
I   *So, you search for sounds to fit the title? Or do you find them by chance?*
AVG No, it's absolutely not by chance. I have a library of my own sounds, and if I need something for a project, I will perform the sounds and capture them myself. Sometimes, I might use samples of music in specific projects, but I usually use only my own sounds.

Expression and Performance

I   *Why do you choose to use only your own sounds?*
AVG Composition is not just about assembly but about expression and performance. So, the very act of capturing sounds is inherent to my process.

I might have wonderfully recorded sounds given to me by excellent sound engineers, but they have no resonance with the project and its concept. If I make the sounds myself, I am able to respond to the nuances and the character of the project idea and make creative decisions within the sound collection process. Capturing my own sounds also gives me access to a context that guides my listening attention. When I record sounds, I have the experience and the memory: I know where, when, and how.

I guard carefully what I like, and I'm not afraid to throw away sounds that do not connect with me. In this sense, I not only create but also curate my own sound universe. And, of course, the characteristics of the sound inform the way in which I am able to process them. Their characteristics dictate their treatment.

I'll give you a little example, in my work *Eau*[89] there is a recording I made of drops of water captured in a cave in the Orkney Islands in Scotland. These sounds are recorded in such a way that introduces colouration, and no sound engineer would choose to capture these sounds like this. But this colouration is due to decisions that I made at the time of recording, in response to being there and hearing the sounds in that location. And my memory of the recording experience gives me a greater understanding of the sound and how it operated in this original space. Later on, in the studio, this colouration informed the way in which I used these sounds as part of the composition process.

Sonic Archetypes

I   Are there any specific characteristics of sound which you are drawn to again and again in composition?

AVG François Bayle[90] talks about sonic archetypes, things that we can immediately recognise and understood by everyone from any culture. So, the archetype is a good tool to communicate meaning directly. There are two main types – the first is entirely physical.

For example, if you take a wave sound and loop it so that it flows back and forth, it conjures notions of rocking – perhaps as you might cradle and soothe a baby back and forth in your arms – this is an impression quite distinct from others such as bouncing or friction, scrubbing or shaking, each distinct physical archetypes of sound.

The second archetype relates to the sonic image, sounds that have a referential relationship with what we know already. For example, everyone will recognise a bird, and when you put birds in your music you perhaps conjure an image in the audience's imagination. Perhaps you get a sense of the openness and the blue (or grey!) of the sky, and this evokes a feeling. Archetypes give you a way to communicate to the imagination of the public. (You can find details about the various categories of archetypes in my book – *Treatise on writing acousmatic music on fixed media*).[91]

In traditional composition, I only ever learned about the architecture of sounds and the work, never about the architecture of the listeners' perception, and this is a big difference!

Thinking about the listener changes the way you think about and approach composing. When I was first studying in Paris, François Bayle revealed to me that he would just use any old settings of the synthesiser to generate sound materials, and it wasn't carefully designed or considered. At first, I was shocked, but I soon learned that being able to touch the imagination of the listener and structure materials in a way that guides their memories and creates associations was just as important as the sound itself.

I might choose to try and convey a particular sensation of a season, and in doing this, the goal is not to choose everything, but a particular moment, which the listener can understand immediately. And I don't mean this in the intellectual sense, but the feeling.

Composer as the First Listener

I   *Is it a desire to communicate impressions of feeling that guides you in the studio?*

AVG   I feel that I am the first listener. I'm not even really a Composer at this point; I am a listener who responds to what the sounds give to me. And from the very beginning, it's all about the feeling and about what these sounds say to me. If they say nothing, I don't choose them! [laughter].

Afterwards, I become a Composer. But I don't just mean this in the traditional sense of someone who assembles sounds – vertically, horizontally, or alongside each other – creating forms with material that repeats or develops. What interests me is to guide the imagination of the listener. To lead their interior cinema, by triggering memories, inspiring their imagination, and creating a reaction in them.

Building relationships in time and space – what you would traditionally call 'Composition' – is only secondary and is really just an organisation of the materials in order to create a structure of listening for the audience, a structure to connect felt experiences and to inspire imagination.

I've talked about the various strategies to articulate the energies and trajectories of sound with François' Delalande[92] and Bayle, and more details will appear in my new book.[93]

Sound Transformation and Performance

I   *Are there certain techniques and approaches that you are drawn back to using again and again?*

I always try to employ a diversity of processes, even within one piece. And yet, there is always a sense of my own personal style that emerges! Yesterday, I shared the piece that I am about to finish with [composer] Hans Tutschku,[94] and we had an extensive exchange. At one point, he said, "I recognise your style!" But I don't personally know what my own style is!

I like variations in speed, transposition, stretching, and band pass filtering, but the processes don't come from the tool; they are informed by the affordances of the sound. The reason for employing a transformation is not the transformation itself, so it always changes based on the context [laughter].

The only time that I let the tools lead is if I am experimenting with something new and I want to find out what it can do. For example, a couple of weeks ago, I discovered the plug-in suite GRM Tools – Spaces,[95] and I've been experimenting with what it offers, so that I can make a decision on whether or not I want to use it within my next piece.

In terms of tools, the ergonomics are really important. It's important for me to have a gestural approach to control. I'm not interested in just replicating analogue functions and features in the digital, but in finding new ways to control sounds. My colleague Benjamin Thigpen[96] has been working with Max/MSP and Lemur[97] on the iPad to create more intuitive interfaces that can control the parameters of the GRM Tools plug-ins, which allow me to just perform and play with the space.

And that is really important to me, because I was primarily a pianist in my training, and so, I have this strong sense that music comes from the body. And I really need to have gesture. The interfaces that we use to control computers, keyboard, and mouse are really terrible interfaces for gesture.

When I am in the studio, my approach is that I play and I record performances of sound transformations repeatedly. I'll explore many different permutations, and at the end, I'll choose the forms that work best for the piece. The result is that you have a lot of variations, and you can choose elements from this library to go into the piece. But at its heart, it all comes from the body.

I   So, you're improvising and then curating?

AVG Yes, absolutely. And the final choice of which variation I go with will be informed by those relationships that we talked about earlier, and how they fit with the idea of the project. The more you explore, the more you understand. It helps you to really understand and to push the possibilities in the materials.

I   I suppose it comes back to a human sense of gesture and the relationship of you shaping these sounds for the audiences?

AVG The result for me is that the sound is more alive. It's a more tangible real thing.

It's a real shame that the standard computer interfaces of the mouse and the keyboard are so limiting. I don't think it's by chance that engineers have conceived and designed software to reflect the analogue, for example with faders that give you the idea of touching something physical. But sometimes that ideal of gestural, tactile, control is missing.

It's interesting that the DAW has developed so that you read it from left to right like an analogue tape machine, but not vertically like we used to do in the film. We've taken the method used with tape. And I suppose my goal is to try and ensure that those other aspects of the analogue studio, those more tactile gestural elements, are carried across too.

That's the reason why I still give classes in analogue studio composition techniques on our Studer[98] 8-channel and four Studer stereo tape recorders, because that tactility is so important to me as a Composer.

I   Do you think of sounds like characters?

AVG   Not exactly as characters. But what I learned in Musique Concrète[99] from Schaeffer[100] is to understand morphology. I learned the morphology of sounds, and I now teach this to my students. So, I can understand and communicate with sounds via this means. If I hear a sound – which is just a slow fade in and out – it says nothing to me, but if there is a more dramatic gesture – [ssshup-ding] – I know that this is triggering something. There is a certain vocabulary of gestures which can be understood… But I can't explain the musicality behind it [laughter].

Concerts and Curation

As far as I am concerned, Acousmatic Music is for everybody, and audiences don't require a specialist knowledge to appreciate it. We recently did a concert in a garden, and the audience was very different from the usual. We had neighbours attend who had never been to an electroacoustic concert before. It was a new experience for them, but through archetypal images in the music and expression in the spatialisation performance, this new public stayed for the whole two-hour concert! And they made judgements and expressed a sensibility about the different pieces, which was very nuanced. And this is the proof for me that this music has the power to connect with people, even though it is very far from popular music.

I believe that one of the most important roles of the Composer is to communicate with the public as an organiser of concerts. And the most important aspect of this is the curation of works. I am an 'expert' in this music but even I have suffered at festivals where there has been no change and no variation in the programming.

It is a question of respect to the public. When all the works are too similar, there is no value to the audience, and it is simply one homogenous mass. If you poorly curate, you will lose the interest of the public because you are not respecting them as listeners. We have such a powerful potential to communicate through a whole variety of sound practices, and it is in that difference that we find the beauty of listening to and experiencing how people communicate to us in sound.

4.2.3 John Young

John Young (b.1962) is an Electroacoustic Composer exploring the use of sound recording as a creative tool. His works merge sound images of the real world with more abstract sonic materials, inviting listeners into imaginative worlds where familiar objects and environments are given new meanings.

His works have received international acclaim with awards including First Prize for his work *Inner* in the 1996 Stockholm Electronic Arts Award (Sweden) and a First Prize for his radiophonic composition *Ricordiamo Forlì* in the 34th Bourges International Electroacoustic Music and Sonic Art Competition (France, 2007), and in 2022, he won the inaugural Prix Francis-Dhomont at the Akousma Festival in Montréal for *Le Chant en dehors*. John is Professor of Composition in the Institute for Sonic Creativity at De Montfort University, Leicester.

Speaker Key

JY   John Young
I   Interviewer

Music Form as a Journey

JY   I've always thought of a musical form as a journey. But recently, I've begun to think about returning to ideas of indeterminacy, and this seems anti-narrative in a way, perhaps more environmental.

And so, my work seems to emerge from that distinction between a) narrative or journey, and b) the creation of sonic spaces or environments that you feel you are inhabiting.

I   *That's interesting, because a journey through environments perhaps needs to be quite open, and non-prescriptive, so that the audience can project themselves into it. Do you feel like you are thinking about the listener as you create?*

JY   I've always liked Stravinsky's[101] expression that he composed for himself and the hypothetical other. I don't really see myself as producing a product. Of course, you are producing an aesthetic object, but it's also important to acknowledge that I also have the view that composing is a form of personal 'therapy' and a way of connecting myself with the world and others. There is an expressive imperative; I hear things in sound that I want to coordinate and understand in my way, and shape into a form of expression that others might relate to.

This sometimes means thinking about the history of music and using what we have come to understand as a musical journey or a musical form – a musical teleology if you like – and sometimes actually thinking about what is outside of that. Perhaps trying to be a bit bold and subvert those traditions in some way.

Chance and Indeterminacy

I   *Do you feel that there is a conflict here between this notion of the journey and this idea of indeterminacy; the clearly defined path and the more serendipitous chance encounter?*

JY   Perhaps. But recently, I've been trying to re-engage chance within composition with acousmatic materials. For example, I've begun to work with MAX/MSP[102] to trigger sound files with live performers. These sounds are intended to integrate with the material being played on an instrument, but it's really interesting to randomly scroll through those component sound files, to discover combinations of these sounds – my sounds that I've shaped – but which come together in ways that I wouldn't necessarily have thought of.

There was an article by Warren Burt in the journal *Organised Sound*,[103] where he talks about his relationship with indeterminacy and connects it to gambling. He tells a story of how he went to a casino with some friends who worked in the finance industry. They were quite happy to put what he thought were large sums of money on a chance of winning something back, which as a Composer he was horrified by.

But at the same time, he acknowledged that as a Composer he is quite happy to try putting sounds together in ways that his friends working in the finance industry would find unmusical or just not an interesting thing to do. There were two views of chance, an economic view and an aesthetic view, which I think is quite interesting.

His idea was that chance is a useful way of learning about what you can listen to. That's one of the great things about the studio work; you've got this canvas where anything can be shifted around and sliced. Working with computers means that often sounds can be projected back at you that you hadn't anticipated. That can be surprising and extend your aesthetic and technical understanding of sound and the consequences of audio signal processing. But there is a need to get beyond the open possibility of "I could take anything" and try to refine that into what it is you actually might be looking for aesthetically, because I think that there is some sort of aesthetic core that we each have as individuals. But that can be hard to define – in composing I often feel that there is a kind of sound that I'm looking for. However, I don't necessarily know quite what it is or how to get it.

Emotional Connection

I   How do you know when you've found it?

JY   If you have to ask you'll never know [laughter].

But seriously, I tend to just get excited. Your pulse starts racing, or you start sweating a little bit, or you get a buzz, or even goosebumps sometimes. Essentially, it does something that enables me to see past it and towards how it might function in or contribute to a form.

Much like listening to any other composer's music when you get that feeling and think "gosh, it's exciting". For example, when I first heard some of Gérard Grisey's[104] music to me it was so powerful, even revelatory – it was suddenly so obvious that you could evoke so much emotively and formally by mining the interior detail of sound in this way.

But if I think in terms of my own process in the studio, it's rather more difficult to put things into clear perspective.

It is not the same as hearing another composer's music because in that case, you have an objectivity, which means you can step back from the work and start to get a sense of what the underlying principles might be. That is much more difficult with one's own work, I find.

But, it's the excitement of engaging directly with sound in the studio I was trying to describe earlier that somehow translates into processes of selection and a motivation to bring a work into realisation. There's something else that happens when I get well into a piece, which is that I feel the material is under my control with minimal effort … I know its potentials and how I can shape it. That also often tells me when I have reached the end of working on a piece, because adjustments or new elements start to detract and take the work into a direction that no longer 'fits'. That tends to lead to a kind of melancholy, like events that have taken their natural course and are now out of reach, or in the past in some way.

It's sounding very Damascene, isn't it?

Emotional versus Technical and Objective Approaches

I   *In a sense, it's not just about what you hear, it's about what you feel?*

JY   Yes, totally. That's actually largely the way I was taught as an undergraduate. There was very little technique that I was taught about composition, certainly in the studio. So much importance was placed on locating a strong sense of identifying with material that there was not a great emphasis on technique *per se* to the extent that, in some way, there didn't seem like there was that much to be taught. But in retrospect, there was a lot actually.

My main tutor, John Cousins,[105] was always pushing towards an ideal that "it's got to hit you that you are saying something", which is great and fundamentally important. But it wasn't until I had direct contact with Denis Smalley,[106] getting to understand his way of thinking that I could start to see that actually you can and should stand back from the sound – analytically in a cool, clinical way. What are its characteristics? What can you do? Then ultimately stand back from the form and say, "What's right about it? What's wrong about it?" The influence of Denis has also been important not just because of his model of clarity of thought and intent, but because as a composer he also exemplified the importance of locating that feeling for sound and form.

Maybe I'm doing John Cousins a bit of a disservice there because he did actually convey some of those technical analytical ideas, principally by giving insight into his own work in progress but it wasn't his main focus – he stressed more the importance of personal investment in the material.

The Human Connection

I   *So, in terms of the aim of the work, when you talk about a "journey" do you feel like it's an emotional journey that you are taking people on?*

JY   Well, it's a human process, so I want others to experience something of the work that I do. Primarily, I think I am interested mostly in my own sense of what is musically satisfying, I would say. But of course yes, there's a communicative imperative behind artistic work. Otherwise, why would I do it? [laughter].

It's a weird amalgam of, (a) having interest in exploring sound and what I can make, taking gratification from that process of sound manipulation and transformation, and then, (b) wanting others to share in that experience of sound.

It might seem a strange analogy but, in the last year, I've started running. I had never quite connected with the idea of endorphins and the buzz that people say they get from exercise. But since running myself I totally understand it, and I've realised that it's actually a very, very similar feeling to an intensive day in the studio.

I don't even have to come out of the studio with anything that I'm particularly happy with yet, with concentrated effort, I still feel very satisfied at the end of the day. It's that same feeling of having exercised and it comes with a sense of positive well-being.

Whereas if I have a day where I'm sitting doing lots and lots of emails that are going nowhere and going back and forward, I come home and feel totally frustrated and rather tense [laughter].

So, I suppose with that buzz I get from composition I'm serving myself and my own sense of well-being, which is a luxury. But I hope that the sounds that I compose do offer an aesthetic experience to listeners, some insight into our experience of sounds.

But, of course, I have also made pieces dealing with oral histories, particularly experiences of war and conflict. So, I think through these more programmatic choices, I can turn some of that more abstract aesthetic intent into something that is communicative about a significant topic.

Experience Relived and Tactile

I   *Yes. It strikes me that those works, Ricordiamo Forlì*[107] *Lamentations,*[108] *Once He Was a Gunner,*[109] *etc., while they come from an almost biographical basis in your parents' experience, there is a wider message that is being communicated through those works about the impacts of war and conflict.*

JY   Yes, I hope so. I mean it really comes from my first encounters with sound recording, and the idea that recorded sound is an experience relived. I am also inspired by the frailties and characteristics of the technology and how they capture traces of the past. So, 1970s cassette recording is quite charming in its way – one can somehow read into lo-fi sound something of the impermanence and potential vagueness of memory.

I don't know if you recall the phonautograph?[110] This very early tool for sound recording, which had a stylus that etched a sound wave in a charcoal surface. That wasn't sound, because it couldn't be played back at the time, but recently researchers in Berkley, California, took some of the earliest recordings and scanned and sonified them,[111] and the fact that this recording etched in charcoal from around 1860 can be turned back into sound actually is really exciting, because sound is alive, isn't it? It's tactile and it's living, so even in a recorded state, it offers a reengagement with that tactile experience.

So that was the idea that I carried with me when I first went to Forlì, trying to understand the stories and the experience that both my parents, but particularly of my mother who lost her sister and a niece in the bombing of the town as well as a nephew in a shooting incident during the Second World War. I was just trying to understand what that might have felt like. But always with the sense that it has a wider significance because similar experiences are still unfolding for people all over the world.

The Sound of Feelings

I   *I think this concept of experience relived is really interesting. Through capturing and composing sounds you can bring to life an experience that reflects memories and exchanges. And sound recordings, in and of themselves, carry a trace of the past in them.*

JY   Yes, exactly. Both sounds and the voice, because in an oral history recording, the way someone actually expresses a view in speech, where their speech might falter as they are bringing memories together, is a reflection of the synthesis that's going on in their head as they recount the story and you can actually experience something of that thought process in listening to their voice. Whereas if you just have a transcript of their words, it's not at all the same thing, it's a reduction.

So, I suppose what I've tried to do in *Ricordiamo Forlì* is use musical sound as a way of aestheticising, trying to re-elaborate the experience of what is going on inside the person when they are speaking. The music is aiming to stand in for the memory that is underpinning the process of forming words, because when we speak, we are often thinking many different things that we are trying to channel into a statement.

That's what I'm trying to do with sound. But not just sound, musical sound. Sound that isn't just 'itself.' Sound that has some relational connection to other sounds, so that there is a wider architecture. The sounds aren't just themselves. They are stand-ins for something other... They have a value or a function somehow, not just illustrative but relational.

Storytelling as a Structural Approach – Trajectory

I   *I think that nicely explains your idea of the experience relived and remediated. That articulating the unsaid elements is really the goal of your music. It's the 'sound of feelings' that's being articulated.*

JY   I think so. Yes, and that's also why I'm attracted directly to the storytelling approach, to get what I feel is a true sense of musicality out of sound when there is no tonality. Tonality is the great creation of many human cultures, the rootedness, and sense of moving towards a particular goal, or not, which is grammatical and syntactical.

Yet, I sense that a lot of electroacoustic music, a lot of different music, doesn't engage in that way, and for me, that sense of trajectory is really important. [Iannis] Xenakis'[112] music is often very successful in that sense, even though it's not tonal, because you are hearing patterns and you are hearing tendencies. You are hearing things moving towards implied states and so on. That's what I understand by Grisey's idea of 'pre-hearing' – some kind of expectation that leads to closed or open states, or tension or resolution, and so on, which is important, I think, to encouraging some involvement in a musical experience.

But I'm then interested in this extra level where you can introduce words and a story, which you can then illustrate musically. And then maybe by illustrating it, you can also direct or construct some sort of argument or dialogue between these interweaving trajectories.

Composition as Poetry – Essentialised Essences

I   *I wonder whether the notion of poetry is useful? Poetry is made of words, but they are always seeking to evoke something else with the language. It's always beyond the words themselves. Would you describe your work as poetic in that kind of sense?*

JY   Yes, I think so. Because it is typically very condensed and essential, it's about essences, which I think should leave you with a wider resonance. And I think that's the sort of thing we can do with electroacoustic sounds. That is what *I am trying* to do with electroacoustic sounds, anyway [laughter].

I remember hearing a film of a masterclass with John Williams, the Australian guitarist.[113] He was extolling the virtues of the guitar and saying, "This wonderful instrument. You pluck a note and it just takes you into silence". And if I think about my instrument, the Acousmonium[114] or loudspeaker orchestra, there is something wonderful about that space in which acousmatic music takes place. I don't mean the space in the speakers. I mean just the space in which you hear it, in which a sound just disappears.

It comes from absolutely nothing and disappears into nothing. That you can do that is just really beautiful in itself. The vacant ontology of music without performers, without the visual. Yes, there are loudspeakers to see, but they are just a channel. You are not witnessing any action.

And then you extend that with collective listening in the concert space, and you make possible a shared sense of moments and experiences that are seemingly unmediated by performers acting sounds on your behalf, but you enact the sounds yourself through listening, it's just really exciting.

For me, that's what is important. I am not so interested in describing the process that I did as X, Y, or Z to get that sound. Well…. of course there is the geeky side of me that is interested in those techniques and that process. But it's not a case of the end justifying the means, or the means justifying the end.

It is always about that end sense of the sound, and everything I do is driven by a personal aesthetic led by what excites me sonically.

4.2.4 Trevor Wishart

Trevor Wishart (b.1946) is a Composer and Free-improvising Vocal Performer, working primarily with the human voice. Wishart has written extensively on the topics of sonic art, composition with sound, and the morphology and psychoacoustics of sound transformations, including seminal books *On Sonic Art*[115] and *Audible Design*.[116] He has contributed to the design and application of software tools used for the creation of digital music – the Soundloom and the Composers' Desktop Project.[117] He has received a variety of awards including a Euphonie d'Or at Bourges and the Golden Nica for Computer Music at Linz Ars Electronica. In 2008, he was awarded the Gigaherz Grand Prize at ZKM, in recognition of his life's work.

Speaker Key

TW   Trevor Wishart
I   Interviewer

The Voice – A Rich Affective Source

I   *In your work, it seems you are always drawn back to the voice. Is there a particular characteristic of the voice that draws you back again and again?*

TW   I tend to divide things into the technical and the poetic, and by poetic, I don't mean to do with words. I mean something that's not technical [laughter].

The technical reason for working with the voice is that it's a very rich sound source. It shifts its spectral characteristics from moment to moment, which makes it both technically difficult and technically interesting to work with as an artist, so the material is very rich and exciting. In contrast, instruments are typically designed to have a fairly stable spectrum, they might have different versions of this, like pizzicato[118] and arco[119] articulations etc., but the whole point of a clarinet is that it sounds like a clarinet, while a voice can imitate almost anything you can hear, and you don't need or have to retain that connection with the source, you don't need to know it's a voice. It might become someone imitating a train, or a dog, or something other.

There's also this slightly different reason. If you make a multiphonic[120] on a clarinet, lots of clarinet players, and people who are really into contemporary music, will be quite interested. But if you make a multiphonic with a voice everybody is interested. They all want to know, "How is that happening? What does it mean? Is that person frightened? Why is that so scary?" We have this very, very direct link with the voice like we have with faces in the visual domain.

We recognise faces where there are no faces at all, all you need is a couple of dots and a line and we can see a face, and that's the same thing with the voice. For example, if you reproduce a voice with one bit resolution rather than 16 bits, you can still tell it's a voice.

So, there's two things, it's technical challenge and richness, but also this link with a very wide audience. People might not like what you do, but they will be affected by it.

Expressive Content – From Micro to Macro

I   *Do you find there's a shift when you're working with actual words in a more narrative sense? I'm interested in how that shifts the way in which you negotiate the material?*

TW Over the last 10–15 years my compositions have taken me from working with tiny fragments of voice-generated sounds, in *Tongues of Fire*[121] – where they could be [makes a noise] noise, you know, any kind of extended vocal technique – to working with syllables in *Globalalia*[122] and finally to whole phrases with the piece *Encounters in the Republic of Heaven*.[123]

When I came to *Encounters in the Republic of Heaven*, I decided to move onto the level of vocal phrases, looking for the musical content within them – for example, they have a pitch contour, they have a rhythm, some expressive or dynamic contour – and these became the musical motif, or a "tune", which you could play with and contort.

But suddenly, I couldn't ignore what people were saying. If you're recording isolated syllables, that doesn't matter, you can just treat the objects as sound objects. But with the complete phrases, I felt a responsibility not to mess unduly with what the speakers had said.

A lot of people from County Durham volunteered to be recorded as part of the project and most of them had no experience of electroacoustic music or possibly even contemporary classical music of any kind. And so, I had to be very mindful of the social dynamic involved, I couldn't distort what they said or make them sound ridiculous, so I had to confine myself to working with parameters like the pitch contour and the rhythmic sense or even the hesitations and the glossolalia in the voice.

I initially did what I call "radio editing", cleaning up the recordings without doing the speaker an injustice, e.g., cutting out repetitions and any uninteresting hesitations. I also did some slight reordering of things so that certain key phrases were repeated like you would do if you were writing poetry. But what I was really concerned with was the way it was said, that was what the piece was about.

That's not to say that I'm not interested at all in what people say; on the contrary, that interacts with what I'm doing; and it perhaps suggests the ways in which I might treat some of the sounds. But I'm primarily interested in the musical qualities, I'm definitely not setting words as lyrics, and I'm looking at the actual expressive content of the syllables or the vocal contour within the spoken voice itself.

Utterances, Glossolalia

I   *I'm interested in your use of the articulations and even pauses in people's speaking voice, because one of the things that has emerged from our discussions with Nina Hartstone (and others) is how those subtle non-linguistic elements are massively important elements for their storytelling – breath, pauses, and the clicking of lips.*

TW *Tongues of Fire* was really about those kinds of sounds – 'glossolalia' – and all the interesting musical qualities of these sound objects. But of course, some languages use sounds that we think of as glossolalia as meaningful phonemes.

I   *Of course, we often default to thinking just about the English language. I've been doing a piece in ancient Babylonian recently, and it's quite interesting to learn about some of the glottal stops that they use, where they really truncate syllables to create gestures and shapes which totally change the meaning of the phrase.*[124]

TW Yes, you should try Vietnamese if you are interested in this. They have a fascinating use of stops and other things. For *Globalalia*, I collected material from all kinds of languages, about 26 languages: Germanic languages for consonant clusters, tonal shapes with stops from Vietnamese, all sorts of vocal gestures that reflect the diversity of languages from across the globe.

What is really interesting to me is that you can speak English without actually having any expressive inflection at all, but you can't speak German like that. You certainly can't speak Spanish or Japanese like that. In Spanish and Japanese, the vowels are very differentiated, but in English you can keep them all in the middle of the mouth, not really moving your lips [demonstrates], and you can still understand what I'm talking about. There's a really good book called *The Sounds of the World's Languages*,[125] which I'd recommend. It's just so amazing. They have a huge chapter on all those vocal clicks you get in Southern African languages and the combinations of consonants with vocal clicks or clicks at the same time, which we find really quite difficult to come to terms with, because we're not accustomed to them, but for the people who live there, that's just completely normal.

Crafting a Structure

I   *If we come back to the question of creativity, then how do you know when you've got an idea that's worth growing into a composition?*

TW   Two things have to come together. One is a formal idea, essentially a musical notion, for example, 'making a piece using syllables', as in my work *Globalalia*. It's usually a very, very vague notion. Then, I set about collecting a lot of materials and begin to work with them to discover whether or not it's going to work! [Laughter].

I   *So, the idea gives you a framework?*

TW   No, the idea is very much a speculative thing. For example, "Can you make a piece where you have theatrical scenarios, and you transform them gradually?" It's not really a framework, it's a very vague notion of the sort of thing you might do. Then you work through the implications, "what theatrical scenarios, what's it about?", "Is it about the state of the world?" "What kind of things would I put in it", or "What possible things can I do?".

There are practical constraints as well, especially so in collaborations, "What will the writer make of it?", and "How will we relate with each other?", and so on. And then, we might be recording the actors and some of the things they do are fantastic and some of them are not so interesting, but that is where the process of editing comes in.

With software, I can modify it, I can make their speech more emphatic – to a certain extent – or speed it up or slow it down. Or even change what they say. There are all kinds of things that I can do with it, and then, I can layer it with other things which can completely change what it appears to be.

Music Is the Impact of Its Formal Structure

I   *I suppose the challenge is that many of our dialogues fall back on the technical and the more analytical reflections on the process because that's a bit easier to explain or discuss.*

TW   I think it's a very difficult thing to explain, even when you're teaching music. Music works through formal structures. Everything that works in music, works for some formal reason which you can tease out. But that's not what the music 'is'. Music is the impact of that formal structure, and we still don't fully understand how that works.

We have a huge history and knowledge of things that work from studying music of the past and things that continue to work if we modify them somewhat. But if you're teaching students how to make a fugue, you're not *just* following these rules. A successful fugue has logic combined with its detailed emotional effect from moment to moment. That is what makes it something powerful musically which we can't explain, and you can't teach it to anyone.

And the other important thing is that you must not assume that because a fugue follows a logical structure, from A to B to C, that's how the composer put it together. They probably started with another subject which didn't work and so they moved on to a second subject, and they tried various things and then they discovered a good bit which they put near the end, and then they started to think about a structure and put together the whole thing. So, when it's finished, it all has this feel of being logically inevitable, but that's not how you make music. That's not the process of creativity.

The Unfolding Process of Composition

I   *I suppose it comes down to the details and the expressions that you put in, you have an overarching formal structure or an idea, and then, the music is actually about that unfolding in time, which hopefully carries the listener?*

TW   What tends to happen in my experience is that I might have certain general notions of the shape of the piece, or how tensions come and go within a piece. But as you start to work with the material on a very small scale, you make little bits and pieces and you build them up and eventually you might make them into an event which is totally spectacular.

Then you think:

> Z is a really important event and it has to go at an important point. I don't know what else is happening in the piece anywhere but I'll put it there.
> Because Z is so important, I should indicate that it's going to happen in some way, so there must be some premonition of this event in the piece, and actually I should do Y.

As a result, you build up these pre-modifying elements that signal to one another. And you have this time structure that has these events in it and nothing else. Then, you might create another event X and think:

> X not quite as important as Z, so I'll put that here.
> But I should prefigure X with another set of premonitions and also ripples after it has passed, so let's do W.

So, as you work with the materials, the form gradually begins to solidify around the materials.

One typical experience that I have when you come to the end is that events which previously seemed well formed now appear too long. On its own the event was wonderful, but in the context of the piece it's just hogging too much time, and it's holding the piece up. So typically, you have to shorten individual events before the work comes together as a satisfactory whole.

I should add that I do all of this by listening. I don't have any formula such as 'golden sections' and all that nonsense. You do it by listening, and you understand these proportions because you experience them.

Towards the end of a work, it's very difficult as you have to listen to the whole duration, perhaps 35 minutes, to make that judgement, thinking "well, that's not quite right. I'll shorten that bit", and then you listen to it again and again.

Emotional Communication

I   *Is it just about formal characteristics, shape, and gesture or is there something more about communication? This question of emotional communication through the sounds.*

TW   You can assume that it's axiomatic[126] that the emotion is taken into account. Music is the organisation of sound in time which has an emotional effect on people.

You can organise sound in time and present it at a computer music conference and people will say, "Oh yes, terribly interesting" but, of course, no one will listen to it again if it's not interesting from a musical point of view. It might be terribly interesting from a technical point of view, it might be a fantastic algorithm, and you might be demonstrating a wonderful thing with it. For example, sonification in science research is a very worthwhile thing to do. It's not music though. It's a mistake to confuse it with music. Whether the music is interesting is something to do with the way that music affects us.

I don't think anyone has a good explanation of how this works. But if you're a musician, you know that that's what it is about. So, the question you ask is sort of irrelevant, if you're not into communicating emotionally, you're not a musician.

I   *I always feel that there is a very human connection in your music, it can be very detailed in terms of the kind of technical articulations that are going on, but there's always this very powerful human element that draws an emotional response from the listener.*

TW   Well, I hope so. I hope I'm doing something right [laughter].

4.2.5 *KMRU*

Joseph Kamaru (b.1997), also known as KMRU, is a Sound Artist, Experimental Ambient Musician, and Producer from Nairobi, Kenya. His works deal with discourses of field recording, improvisation, noise, ambient, radio art, and expansive hypnotic drones. In 2020, he released three albums: *Peel*, an LP of hypnotic drones on institution Editions Mego; *Opaquer*, a collection of sound sculptures; and *Jar*, a cassette of found sounds and analogue ambience. Along with other compilations and self-released EPs, he has earned international acclaim from his performances.

Speaker Key

I   Interviewer
KM   KMRU

Sonic Environments

I   *What drives you to choose the soundscapes that you use when you're working. What is it that attracts you to certain soundscapes instead of others?*

KM   I'm usually thinking of the space overall, rather than individual sounds that I might get out of the space. When I'm in a particular environment, I'm not trying to look for specific sounds that I can record; instead, I'll often leave the device running to capture a long take and allow myself to be part of the recording as I listen and move around, being drawn to different features of the soundscape.

     It's more an intuitive way of studying a place. I'll make a recording as I move through a space and it's only when listening back later that I get a sense of what specific components of the soundscape might inspire creativity.

     Only rarely will I have a project that centres around a specific sound as a goal.

Space as Relationship

I   *I'm really interested in this notion of the space. Are you thinking about space as a box within which things are contained? Or is space something that's made out of the relationship between events and things?*

KM   It's more the relationship of things that are happening within the site and not so much the acoustic phenomena. For example, I did a project in Kenya in the Kibera slum, and this location was totally different from anywhere I had recorded before. I spent my time just being in this new sonic landscape, trying to understand what's happening in it by immersing myself in the location, which of course also includes the actions of all of the people who are inhabiting this place. The character of the space was more important to me than any individual sounds that I thought I was going to get from it.

I   *What approaches do you use to guide your recording?*

KM   I always try to visit the place before I make my trip to record. I'll go without recording equipment so that I just listen and get a feel for what sounds might be there and what sound events might happen. I'll try to have a conversation with somebody who lives there, as they have a much deeper relationship with the place and know the sounds much more intimately than myself. This gives me an overview, or synopsis, of what interesting sound scenes unfold in this specific place.

Only when I feel comfortable with the place will I go back and intentionally record. Having this familiarity with the location and an understanding of the dominant soundscapes makes you much more attuned to the space when you return. As a result, when you go back to record, you can almost feel like you're a part of the environment. You understand the choreography and the flow of the soundscapes, and you can move fluidly around to collect them. It's a much better situation than going into a place and hitting record without really knowing what will happen.

I   *Where did this awareness of needing to do this research come from?*

KM   I bought my first recorder in Kenya and began experimenting with recording soundscapes. It was just so intriguing to discover how much sound there was around me and I used to record so much and go into different locations. I was capturing sounds and archiving and collecting them, but I wasn't yet composing.

It was when I was invited for this project in Kibera that things changed. I stayed for two weeks collecting sounds and understanding the people involved in this slum, and it gave me a different way of thinking about the sounds that I'm recording. I wasn't just coming in as a visitor, recording sounds and then leaving, and I also realised that I didn't need to be recording all the time. I became aware that it's more important to just listen and be attuned to the specific place before you even get to the recording part.

Connection with the Landscape Through Listening

I   *You want to respond sensitively to the experience of the people that live in this space?*

KM   Yes. I've been making an effort to go back to sites and places that I've recorded before, and to listen out for whether something has changed. You're connecting yourself with this landscape as a human being, through listening.

I   *Have you had interesting reactions from people when you played back pieces and recordings to them of their space?*

KM   I was part of an exhibition in Nairobi and my piece involved lots of tout[127] calls – inviting people into the matatus,[128] into the buses, etc. – and conversations from the street and on public transport. Most of the visitors to the exhibition were intrigued because they usually only hear these sounds through daily encounters, and so hearing these sounds in the context of art was a revelation to them.

Another example comes from when I was recording, a guy came to me and asked me what I was doing. I simply gave him my headphones as a reply, an invitation to listen, and he was so blown away by how the microphone could really amplify the environment.

Evolving Soundscapes and Chance Encounters

I   *You mentioned this idea of change, and how soundscapes and sound environments change over time. Do you think change is an important part of your work?*

KM   Recently I've been thinking about the environment just outside my window to inform my music compositions. You quickly realise that soundscapes are ever evolving, changing, and very spontaneous. Things happen organically and in an uncontrolled way. So, I've been trying to borrow and implement this in my process of making, so the tools that I'm now using include random or spontaneous functions and if I make mistakes or abstract changes sometimes I decide that it is better to embrace and keep these 'accidents' in the work.

When I moved to Berlin, it was quite a shock how different the soundscape was to that of Nairobi, it's much more silent here. I spent a lot of time just listening, to become more attuned to what's happening here, to welcome new sounds into my sonic vocabulary. I was looking for 'noisy' sounds – street noises, human voices, dogs barking – because that's what you get in Nairobi. So, it's been fascinating to uncover the relationships, and differences, in sounds which change as you travel and live in different places.

Creating Real and Imagined Soundscapes

I   *A couple of months ago I attended a talk that you gave where you spoke of moving from the city centre to the suburbs while you were growing up and how you were also very aware of this change in the soundscape. And again, you mention this move to Berlin and how that also changes the environment and creates a different feeling. Do contrast and comparison play a part in your works?*

KM   When I'm making compositions, I'll generally use the recordings from one place, overlaying and merging them together to create new environments and hyperreal soundscapes.

But if I'm creating an installation where there is more space and people can move around, then I'm more interested in highlighting the difference between sites and provoking the listener to listen to different soundscapes.

I   *When you talk about sounds happening around you and sounds existing, it makes me think of the John Cage interview where he says, there's a difference between music and the sounds of the street outside. Music is someone trying to communicate, while sounds are just acting.[129] Do you think that that's a distinction that reflects your music?*

KM   My process is very intuitive and I'm always listening to discover what is happening when I bring different soundscapes together. I'm interested in creating new environments of sound.

I   *Do you seek to create a specific experience that you want that listener to hear? Or are you creating a new sound environment that people can venture into?*

KM   It's an individual experience for each listener and I'm not very focused on one specific intention that I want the listener to perceive from the piece of music. I'm always keen to leave it open and simply invite the listener in.

Sometimes, I'll include a score for the listener to use while listening to the piece, but always keeping it very open to what's happening.

It may be relevant to share that I've been doing listening walks every day without a microphone. Instead, I have been sketching graphic scores that reflect what's happening in the environment around me.

I   *So, it is a form of recording? But an impressionistic one?*

KM   Yes, it's this idea of trying to visualise the landscape in whatever form that I understand, not simply archiving the space, but trying to make sense of it. I could use this to make instructions for a listener or even use it as a score if somebody wants to perform this environment.

I   *I really love this idea that you might hand this score of the soundscape to an impro-*
       vising jazz trio[130] and see what they bring out of it?! What is it in particular about
       the soundscape that draws you in? What is it about the sounds that draw you in and
       why do you feel that you engage with them?

There Are No 'Good' or 'Bad' Sounds

KM   I've realised that, for me personally, there are no good or bad sounds when I'm out-
       side. I'm more interested in what is the most interesting sound.
           If there is a sound that is a nuisance or has higher noise level, then I have to em-
       brace it and make my ear welcome it. These sounds are all part of the environment.
           If I'm on the train, I may be intrigued by a specific sound that's happening or
       just the immediate soundscape that's around me. I'm trying to listen to the rhythms
       of what's happening around me, and by rhythm, I'm not speaking about the beat
       patterns of a rhythmic sound, but more about how the soundscape unfolds – when
       people come in, how people sit down, the train is moving, there's sound happening –
       and this whole collage of soundscapes is all interacting with itself. It's very organic
       and it just interests me how things unfold. And how we are a fundamental part of
       it and can influence what's happening. It's as if every person is a performer, they all
       have a sonic role in the soundscape but they are perhaps not aware of it. The more
       I've been listening and recording, the more I'm aware how we influence what's
       happening around us and I'm always trying to be a good soundperson in a specific
       sound place.
I   *It's very easy for us to think of ourselves as detached listeners or observers, but your*
       presence in a space makes you an active participant in that environment.
KM   Absolutely, yes.
I   *How important are the locations and places that you explore?*
KM   I'm not interested in coming in as a random person, capturing the sounds and leav-
       ing. It takes me time to negotiate and explore a space. I'm keen to ensure that I'm
       not appropriating anything.
           I have recordings that I did three years ago, and I'm only now revisiting these to
       edit and work them into an installation. It took me this time just to process this place
       and to understand it. I'm not interested in splicing up recordings and using sound
       fragments, but more thinking about the place itself.

Giving Voice to Objects

I   *I think it's a really interesting balance, as a Composer, to not impose yourself over*
       things, to make sure that you can let the sounds themselves and the spaces them-
       selves speak. It's almost like you have to curate, rather than compose?
KM   Yes. I think that the things around us have voices, or want to voice something, and
       so my role as an artist is to enable them to speak. How can I take this glass [picks up
       water glass from the table] and make a piece for it by allowing it to share all of the
       sounds it can make.
I   *To what extent is duration important in your practice?*
KM   It is true that I have tended to make long-form compositions, because I believe that
       it draws the listener in. Knowing that the work is 15 or 20 minutes long allows the
       listener to prepare themselves for a fuller and deeper engagement.

I used to make much more rhythmic electronic music with drums and play this in venues and clubs, but at a certain point, I wanted to slow things down and allow people to really listen in amongst all this chaos that's happening around us. I found myself leaning towards slow and ever-evolving changes, with fragments of field recording. I couldn't do this in a three-minute piece because there simply wasn't enough time for things to unfold.

I   *Your framing of sounds seems non-hierarchical, and you're not making value judgements about good or bad sounds?*

KM   Yes. And not trying to conceive listening very narrowly in an anthropocentric way of thinking, but understanding how our world is very polyphonic.

I think people become more aware of the 'other' when we become good listeners.

Field Recording as a Fluid Process of Discovery

I   *Yes. And that ties back to this point of you being just another object within the soundscape. You're not in a privileged position; you are simply an active participant in a world that is unfolding around you. I feel that feeds into the way that you think about creating your pieces, creating spaces for listening.*

KM   It's focusing on individual experiences in relation to the sounds in the space that are happening, letting every invited person to experience the piece based on their own memories or thoughts.

The last piece I made, *NUBI*, was in a dark room obscuring the ocular view of what's happening and letting people move through their bodies in relation to the sounds that are happening.[131]

I   *Would you say that your works are quite open?*

KM   Yes. I might have a narrative that I want to portray based on the specific location in which I recorded the sounds, but when it's out there, it's for people to interpret the piece in relation to their own listening context. Leaving it very open minded in how one wants to relate to the piece itself.

I   *Yes, so it always comes back to listening and exploring that environment and it's a process of discovery, do you think?*

KM   I think field recording feels very open in itself. The microphone extends your hearing and gives access to a huge full spectrum of possibilities.

4.2.6 Annie Mahtani

Annie Mahtani (b.1981) is an Electroacoustic Composer, Sound Artist, and Performer based in Birmingham, UK. She has worked extensively with dance, theatre, and on site-specific installations. Annie has a strong interest in field recording, and her work explores environmental sound, abstract and recognisable sound worlds. Annie works with multi-channel and ambisonic audio in fixed medium works and live performance, and her music has been recognised and performed extensively in concerts and festivals internationally.

Speaker Key

AM   Annie Mahtani
I   Interviewer

Sonic Inspirations

I   *What inspires you to create a piece?*
AM   The starting point is always the sound. And it's always sound that I've recorded. Through the recording process I'm listening out for characteristics within material – something that's got the potential to inspire – and it's this which triggers the piece.

       Once I have the sounds, I formulate a concept that builds on them. This may drive me to seek some other external inspiration – perhaps something that's associated with where that sound came from, or the experience of collecting that sound?
I   *Is it important to you that it's a sound you have recorded personally?*
AM   Yes, it always has been. I've always been very purist about that. It is in listening to sounds that it all starts for me. That listening, in the moment, helps me to decide when and what to record. So, I have a relationship with the recordings from the moment that I hit record.

       When recording, I'm always thinking about composition as well. I'm thinking about sonic space – where I place my microphones, how close I want to be to tune into a particular sound versus how much of the wider space I want to capture.

       However, if I'm working on a sound design project (maybe with a dance company, for example), and I need to turn things around a lot quicker, sometimes I don't have the option to go and record sounds in this way. For instance, the recordings that I could get might not depict the right atmosphere that we need to create for the piece. So, I have to source things elsewhere.

       But if we are talking about my own compositional practice, something less pre-scribed, that's all about me and, that's why I'm so belligerent about it, I suppose.
I   It's interesting that you describe *a more pragmatic approach for what you call 'sound design' projects. To what extent do you think that this desire to work only with your own sounds in your composition work is a purely dogmatic or ideological approach?*
AM   I suppose you could say that it was ideological, but I think it is more than that, because I consider the recording session to be part of the composition.

       A successful recording is defined by my compositional desires; these are what inform the choices that I make when recording, whether that is within a space or within a studio.

       I have a large sound library, full of sounds that I've personally recorded, and that is really important to me. Sometimes, I might discover halfway through a composition project that I need the sounds to be different in some way, in which case I might go back out to record again.

And likewise, I wouldn't share my recordings, they are not part of 'stock sound libraries',[132] they're mine. And so, unless it was part of a collaborative project, they're not really accessible to anybody but me. So they are unique in that way.

Sonic Perspective and Space

I   *I'm struck that this is perhaps more to do with listening to space and perspective, than using these to make decisions on microphone placement and how proximate you are to the various sounds that you record?*

AM   Yes, it definitely is. I'm always thinking about perspective when I'm recording. In any session, I will usually make multiple simultaneous recordings of a sound or a space, which allows me to explore contrasting perspectives – from contact microphones[133] to ambisonic microphones.[134]

Listening to find the right place to record from is the most important aspect. It's like taking a photograph, you've got to frame it so as to get everything in perspective and in proportion, so as to get a really full and beautiful sonic image of that space.

And those perspectives of the space determine how I might use the material moving forwards; some sounds might lend themselves more towards ambience, while others might be much more detailed and zoomed in.

I   *It seems to me that you're always thinking about sound as part of a space and an environment where it's resounding.*

AM   If I'm trying to teach my students about space, I will take them to sit on a bench in the middle of campus and encourage them to think about how we perceive sounds in the real world. We think about sounds that are foregrounded and those in the background. Sound trajectories, morphologies, and gestures – sounds that emerge, sustain, and disappear. How sounds move and shift in space and time. All of which we can use as inspiration and as a guide in composition.

For example, in my work *Aeolian*,[135] I wanted to share my experience of extended listening with the audience and, in so doing, encourage deeper listening. The work compresses a 90-minute recording into a 12-minute piece.

I framed excerpts from the main 90-minute recording with hydrophone and contact mic recordings, so as to draw in the listener to hear intricacies and subtleness in the sound. This first section enabled me to open out into the main field recording with the listeners already attuned in terms of their own listening. This meant that the listener can sit in a recording for perhaps longer than you would normally expect and enjoy the subtle changes in the wind and the birds.

I envisaged my compositional goal for this piece as, "How do I take these people on the journey?". I can't possibly say, come and listen to my 90-minute recording, because it's too long and you're never going to experience it in the same way anyway. So, you need to find ways to highlight an essence of what it was.

I   *What appeals to you in using open durational field recordings in your works as opposed to studio recording?*

AM   The reason I do the long durational recordings is because that's the only way to really capture a place. I can record anywhere for ten minutes, but mostly, I'd be recording my impact on an environment, whereas over 90 minutes, the wildlife begins to come back.

I turn to nature a lot to help me in my composing, natural gestures, and natural events; they're all fascinating. And I think that the act of deep durational listening is really important for mind, body, and soul, but also to connect with nature, our ears, and the environment.

I get really excited when I listen back to some of those recordings; they take me back to a very specific place. And in doing so, they trigger ideas that make me want to create a piece.

I   *I'm interested in when the concept for the work emerges. Does the concept come before the process of recording or is it informed by that process of recording?*

AM   It works both ways for me. Sometimes, I've got an idea for a piece and I'll just go out and record. For example, my work *Past Links*[136] is a piece about a place, a community, and a specific space. I had a concept to go and record in the Black Country Living Museum and create a piece in response to this location. But for my work, *Aeolian* I did not have a driving concept. I went on an intensive recording weekend and was inspired by a specific experience that I had. I wanted to share the essence of this experience with my listeners through the piece.

I   *So, each recording has a fundamental relationship with the conceptual framework of the idea?*

AM   That's exactly it. Everything's bundled into a concept, which informs how I process and make sense of my material and make compositional choices.

It's also creating limitations, if I had access to every sound in the world, it would be much more difficult for me to find a starting point. For each composition, I start with a unique set of source materials and I create everything from that pool of sounds. It would be very rare for me to use a recording across multiple different projects. In fact, I really don't want to be using the same sounds in every piece. If I were to recycle things it'd be like sampling myself. I wouldn't want anyone to think, "Here she goes again. That's the only sound she's got".

Sonic Communication

I   *What is interesting about your response is that there's a running thread of a shared listening experience and wanting to communicate to people?*

AM   Music is a form of communication. It always has been for me. So, I am always thinking: why am I writing this piece of music? What is it saying to people?

We can all listen to highly technical stuff and appreciate it for its technicalities, but for me to be able to really get into a piece of music, I need to feel like I am communicating something for the people who are going to listen. Normally, I'd like to think that a person can listen to a piece of music and feel the essence of somebody or something in there. Why would I ask people to listen to a piece of music, otherwise?

I'm always much more moved by something that I feel is taking me on a journey or is clearly saying something to me. A really important part of my composition process is in drawing out the musicality of sounds and getting them to speak in some way.

I   So *what is it that separates composition from design?*

AM   Composition is my voice. For me, design is technical, where there's something very specific that needs to be delivered. It's got a purpose and it's got to serve something very specific.

I've been busy working on a series of sound design projects – for dance,[137] podcast,[138] and theatre piece[139] – all of which require me to design specific gestures or actions that emphasise or provide the context in which we are listening, but also depict mood, framing the listeners' experience to support everything else that's going on. It often ends up being classed as a sound score for the companies.

I   *So, it comes down to this question of agency. Are you in control or is someone else in control? Do you think there's less scope for you to be creative within the design projects? Are you not still creating structures and shapes and forms within the design elements?*

AM   I think there's a different skill set. When it comes to sound design projects, that's when I will draw on more external material. I'm not a professional Sound Designer by any means. I wouldn't know where to start with most things, but I did have quite a fun time designing police radios a couple of weeks ago.

But it's a very different practice from my compositional approach. In the past, I've tried to approach sound design projects with my compositional methods but I soon realise that I don't have the materials and the resources to make this convincing.

When I'm making recordings, they are for a specific composition, so even if I want to use my whole library for a sound design project, I think I would be stuck.

Role of the Listener

I   *To what extent do you think about the audience and the listener in the process of composition?*

AM   Listening is the beginning, the middle, and the end for me. If I'm writing a composition for the Loudspeaker Orchestra, first, I need a concentrated period just focusing on myself and the sounds, allowing myself to become self-obsessed. Then, later, you need to be able to come out and reflect on the work from a more external perspective. I do want the listener to feel something when they listen to my compositions, but it doesn't really matter to me whether they feel exactly what I am feeling.

If I'm doing sound design, it's much more prescriptive; I'm making sure that I am explicitly underscoring the moods and emotions of what's going on stage. So, I'm thinking about the listener a lot more in those kinds of situations where it's really important that I am giving real signals as to what they should be feeling.

I   *So, you feel that there's more openness within your own compositions, whereas the sound design work is more directed?*

AM   Yeah, absolutely. With my own work, I can do what I want with it. The concept is just a starting point. I can veer away from it, as well, if I need to and it's much freer in that sense. I am definitely open to interpretation, absolutely. But with something that is a prescribed project, it's got to deliver in some ways.

4.2.7 Nikos Stavropoulos

Nikos Stavropoulos (Athens, Greece, 1975) is a Composer of predominantly Acousmatic and Mixed Music. He read music at the University of Wales (Bangor), where he studied composition with Andrew Lewis and completed a doctorate at the University of Sheffield under the supervision of Adrian Moore.

His music is performed and broadcast regularly around the world and has been awarded internationally on several occasions. His practice is concerned with notions of tangibility and immersivity in acousmatic experiences and the articulation of acoustic space, in the pursuit of probable aural impossibilities.

Since 2006, he has been a member of the Music, Sound & Performance Group at Leeds Beckett University where he is a professor in Composition and lectures on Electroacoustic Music. He is a founding member of the Echochroma New Music Research Group and a member of the British ElectroAcoustic Network (BEAN) and the Hellenic Electroacoustic Music Composers Association (HELMCA).

Speaker Key

I   Interviewer
NS   Nikos Stavropoulos

Gesture, Reality, Feeling

I   *In your practice, you're often working with sound objects in a performative way, and there seems to be a very human element to your work, even though you're working through abstract sounds. Is this something you're consciously thinking about when composing?*

NS   It is a bit of both. There is no overarching plan, but there are choices made at every step of the way. And these choices often have to do with the idea of physicality, in gesture, but also in the type of materials and textures. I'm keen that the action feels believable, and as real as possible, because I find that makes the flow of the musical text more successful. This same approach applies to the processing, the assembly of materials, and, more recently, the recording stage as well.

I   *When you talk about reality, are you talking about representation, or are you talking about materiality?*

NS   It's not about representation and it's not about identifying a source, it's about the materiality of sound, how that is perceived, and how that is felt. It's about the tangibility and presence of sound. That's what I mean by reality, rather than the representation.

I   *Do you think that emotion is an important part of your practice?*

NS   Emotion is a difficult word because we usually use the word to refer to specific feelings that we share. I have been working with no concept[140] at all for a long time now and I defend that. It is important to me. I find that concepts may hinder the results. You often work *for* the concepts and not for the experience. And for me, it's all about the experience. This starts with me in the studio, creating experiences that are captivating, real, and tangible. They are engaging, as a result of being real and tangible.

I   *To what extent is the audience an important part of your consideration in practice?*

NS   I don't think about the audience much other than to imagine that I'm dealing with creatures like me, fellow human beings. The idea is that if I, as a human, experienced it in a certain way, they, as human beings, will also experience something that is very similar. More often than not this seems to work.

   People shy away from talking to others about their art, but when they do to me, they often refer to exactly the things that I have had in mind when composing. Just as you did. We haven't ever talked about my practice before, but you said, "Your music is about performance, tactility and texture". And, for that to be clearly communicated through the work is very important to me. But I don't compose for a specific audience, I compose for people.

I *So, your practice is always driven by your response to the materials. And I suppose, it's necessary to have that understanding and that trust in the materials, that they will carry you where they need to go?*

NS There are factors one does not necessarily start with, but which appear along the way. I'm seeking to create a reality, or a hybrid reality, through physicality and experience. You have the material, you have the narrative and then there's also the form, and how the two come together.

Narrative and Musical Dialogues

I *Could you expand on what you mean by narratives in this sense?*

NS We're talking about a more abstract narrative that doesn't relate to language, but does relate to a sequence of events that our sonic characters are engaged in. The structure of the sound material needs to make sense and needs to be believable. That's why I like to work with sounds that give you the feeling of tangibility; this accommodates the flow of events and makes it more believable.

I *This dialogue between sound materials is really clear in your work, and there is a sense that everything has a purpose when it unfolds, all in relation to the context that you have established.*

NS That's very important to me. And I work very, very hard on that. Sometimes, it takes hours and hours to go through a few seconds because it's not working as I feel it should and I reflect a lot on that, creating a continuity that is within context and makes sense.

I *Visceral is a word that I would use to describe your pieces, and sometimes, they can become very, very rich in their layering. To what extent are you thinking about questions of density and 'masses of sound' as a Composer?*

NS When there is layering, I always make sure that the layers are talking to each other, they're reacting to each other, they're touching each other, and that emphasises the action. Of course, it also has to do with the complexity of the narrative, and the agents and characters are reacting and interacting.

I *There's a dynamic in your work from the very, very small – almost granular – to larger clusters. I'm just interested in how you sustain those individual narratives and characters within larger masses of sound.*

NS That is perhaps the trickiest part, one must have prepared the stage in a way that they appear at the right time and for the right reasons. It's what Dennis Smalley called 'Finding' the right psychological space for the sound.[141] The more diverse the materials you're introducing, the harder it is to do that. My view is that the more diverse the materials you manage to bring together in a narrative, the more effective and satisfying the narrative is.

I   *People talk about individual sound objects and individual sound events, but these kinds of conglomerations of sound when everything comes together that's not much talked about, I suppose because it's highly complex, and related to the context that has been established in advance and that is unfolding?*

NS   The compositional reality of sound objects, and/or sound events, is that they are more often than not, very composite things. These might appear as a single sound behaviour, which is easy to refer to and talk about. To discuss the inner complexity of those masses, their construction, and how they relate to each other is much trickier indeed!

Spaces, Proximity, and Intimacy

I   *Do you think you're drawn to certain types of sound material?*

NS   I've been making the same piece for 20 years [laughter]. Sometimes it gets a little bit better, and sometimes, it's a step back. But it really has to do with what kind of sounds display a heightened impression of physicality. Of course, there's a lot of granulation[142] and granular reconstruction going on. But I'm consistently drawn to sounds that have a high-frequency content and transients.
Sounds that are present and that appear intimate.

I   *Proximity is key then? This feeling that you are close to the sound materials makes them pleasant?*

NS   Smalley uses Edward Hall's proxemic classification of distances – social space, private space, personal space, and intimate space,[143] very effectively to discuss how we interact with sound in this context. Music rarely comes into what we perceive as intimate space, but it is very exciting when it does.
Proximity is what happens physically, and intimacy is what happens psychologically. And I do prefer sounds that display those characteristics. I use them a lot.

I   *I suppose this relates back to your point about the experience of listening and how you listen differently to sounds that are in that intimate space versus how you listen to sounds that are in the kind of social or public space.*

NS   It's just a reaction to the material. The spectral quality of materials plays an important role in that, but also the morphology and how the spectrum changes over time. But, it's not something that I'm analytically deconstructing. I'm working intuitively through listening to the sounds that I'm working with.

I   *I really like how you make this distinction between proximity and intimacy, the physical and the affective experience.*

NS   For the last four or five years, I've been working with recording techniques to emphasise this. The hypothesis was that tangibility and presence are improved when you also capture the space that the sound creates as well as the sound itself. I find that this approach also heightens the feeling of intimacy.

Spatial Microphone Recording Approaches

I   *Could you give an example of these kinds of microphone approaches? What are you doing differently now that you weren't doing previously?*

NS   I watched a YouTube video of plastic balls bouncing around in a frying pan, and I tried to imagine how you would capture the aural architecture of that. This led me to a series of experiments, capturing sound with multichannel arrays in very, very small spaces.

We built five-channel arrays with a circumference of 5 cm. Very, very, small. And we got some very interesting results. Then, I took that signal and processed it as a multichannel file. So, the space that you recorded informs the kind of spaces that you get at the end. And it goes through the processing at the same time as any other parameter.

So, it's both about a recording and also how I am working with the recording. I've written a bit about this in an article I published for the Sound and Music Computing conference.[144]

I   *And you applied these techniques to create your piece 'Claustro'?*

NS   *Claustro* is the third in a series of works that make use of these techniques to explore the concept of aural micro spaces. *Topophilia*,[145] *Karst Grotto*,[146] and *Claustro*.[147] The idea is that these are aural architectures which exist, but they're not accessible if they're not mediated by recording. You can't put your head in a frying pan, but maybe you can record that space or something similar. It's the aural architecture.

These are intense spaces, and they can cause phobias, or they can cause philias, for example, like ASMR,[148] which has positive connotations, or claustrophobia, which has the opposite, of things being too close.

I   *So, the space is just as important as the events that are taking place.*

NS   Yes, absolutely. This idea that you have a sound, and then position it in space, is perhaps a very visual centric way of working. In terms of how acoustic space actually works, there is no space where there's no sound. There is no sound event that exists outside of a space. That's not how it works in reality.

Working in this way, I was trying to unify Denis Smalley's ideas of spatiomorphology and spectromorphology in the workflow.[149] The work is spatially very rich, but there's no panning automation because all the space that's created comes from the space that is recorded and the processes that it went through alongside every other parameter. Space wasn't processed separately.

NS   It is a big step. I didn't think it would be, having worked in multichannel formats for such a long time, but using this approach the stage becomes bigger, which creates new impressions, but perhaps more importantly, it allows for a completely different emotional response.

4.2.8 **Natasha Barrett**

Natasha Barrett (1972) is a Composer exploring new technologies and experimental approaches to sound in a broad range of contemporary music, including concert works, public space sound-art installations, and multimedia interactive music. She is internationally renowned for her Electroacoustic and Acousmatic Music, and use of 3D sound technology in composition. Her work is commissioned and performed throughout the world and has received over 20 international awards including the Nordic Council Music Prize, the Giga-Hertz Award (Germany), five prizes and the Euphonie D'Or in the Bourges International Electroacoustic Music Awards (France), two first prizes in the International Rostrum for electroacoustic music, and, most recently, the Thomas Seelig Fixed Media Award for 2023. She collaborates with performers, visual artists, architects, and scientists, and she is also active in performance, education, and research.

Speaker Key

I   Interviewer
NB   Natasha Barrett

Process or Premeditation

I   *When you are making a piece, to what extent do you feel that you are working to a pre-existing plan? Or is it more a process of discovery?*

NB   At the outset, I'll consider, "What am I trying to express, what is the motivation behind the composition?". I may start with a theme, but that might be an emotional expression, or it might be something concrete. If I have free choice, then it will normally be an expression of an emotion or an experience, something I want to communicate to somebody else. But, I rarely know how long the piece is going to be. I don't know how much time I need to communicate the expression, and I might not even know what sound materials I am going to use. These things converge gradually over the months of working, throwing things out and finding things that seem to function. So, the plan starts to 'find itself' through the nature of the materials. The musical structuring decisions emerge during this process. You could also say that the overall plan emerges or converges.

I   *So, there is always an emotional driving force within the piece, or it is always about an experience that you are trying to communicate?*

NB   Emotional expression and communication are both important. But then, you realise that you are making music that other people actually want to listen to! [laughter]. Then, that too becomes important.

I   *Do you find yourself re-using general techniques or approaches that allow you to communicate these expressions and experiences? Have you found commonalities?*

NB   Often, they are unique to each piece, though I realise this is not the most efficient way of working! A visual artist may exhibit a series of pictures on a theme or explore a certain technique. But, if we did that as composers, then we would end up with ten short pieces which that are basically very similar.

I   *So, the uniqueness of the piece is important to the identity of the concept and the idea of each one?*

NB   Yes.

New Tools and Technology

I   *It seems that new possibilities are an important part of what drives you to make pieces?*

NB   I am embracing a lot of technology in my composition. The technology is there as a tool to express ideas. I should not deny there is excitement in the process of simply exploring technology: the tinkering process can be exciting in itself. It is something I love to do, which fortunately is advantageous to my composition.

    I have been composing for a long time, so I have a knowledge of what many sound types and processes have to offer, but of course, the possibilities change because technology changes. I have often found that my ideas go beyond the possibilities of current technology, but then new developments later allow me to achieve these ideas.

I   *Some people create pieces where they start with flat materials (mono or stereo materials) and they build them up into a space, but in a lot of your more recent work, you are using ambisonics to capture three-dimensional spaces. How has this shift into ambisonics shifted the way you approach composition?*

NB   I am drawing on, and exploring, my own spatial experiences of the real world. I want to highlight these spatial features for other people to experience. So, I am still thinking about traditional electroacoustic ideas of textures, gestures, and movements, but manifested as spatial phenomena, exploded as a spatial experience in the music.

    Sound has many facets and meanings, especially when separated from a visual source. It is an acousmatic result that has meaning beyond the spectrum, which I find easier to explore by starting with a complete 3D sound field.

Impressions of Reality

I   *To what extent do considerations of reality impinge on what you are doing musically?*

NB   In a musical context, I find that realism is sometimes better expressed through abstraction. We can never express complete reality through a single recording or single sound because it is just a moment, a tiny snippet of time and space. Then further, who's reality are we thinking about? If reality is an all-encompassing understanding of the environment of which we are a part of, in composition I must do more than explore individual field recordings. I try to engage the material through different degrees of abstraction, which then implies a perspective to which the listener may find a connection beyond the face value of the sound.

    I think it was Michel Redolfi[150] who pointed out that footsteps tell you about the environment – you not only hear the material being stepped on you also hear the acoustic space. I have used a crow or a seagull call in recent works. These are loud sounds that simulate reflections from the landscape and inform the space, which in turn connects to a degree of reality.

I   *So, the bird is a marker for people's experience of spaces, and it is not really about the bird itself. It is about what the bird articulates through its interaction within a space?*

NB   Absolutely. Yes.

I   *So, in this layering of abstract sounds, are you seeking to draw people's attention to the specific elements? Or to what extent is it an open environment that you are creating for people to exist within?*

NB   I have realised that I have to be very careful when thinking about the listener. I do not want to change the way I am working to accommodate listeners who might find my music challenging, yet I want to create an environment for listeners to find a place within.

I   *Do you feel that you are inviting the listener into the work to undertake a process of discovery themselves?*

NB   Yes, this is something I am interested in. My tactic is to create many layers and to gradually expose the features of each, in turn. I present an element from one layer, then remove it and expose something from another layer, and this gradually constructs the different layers. But, the sequence of their unfolding shifts your relationship to each part. I am hoping that listeners can hold onto some of these elements, like threads, and use them to orientate themselves in the materials. They will recognise certain features surfacing, while others sink and get lost. But, this is a fragile situation as you don't know the audience or how they are listening.

Audiences and Listeners

I   *To what extent are you thinking about the audience when you are working?*

NB   I work from my own perspective. The result is that some of my pieces can be challenging for some listeners, but not for others. Sometimes I think that I have made an idea incredibly clear, but when I come back to it after three or four weeks, I realise that it is barely audible [laughter]. So, there is a long, iterative process of balancing clarity without making something blatantly obvious.

I   *Do you think there is also a wider appreciation for sound today?*

NB   I think there is a shift and I think it is a shift that has partly emerged from a complex entertainment and media demography and sociology. Others have studied this much more than I have, so I am no expert, but it is clear that our everyday sound world has changed because of the media surrounding us. Most obvious is the way we are accustomed to listen without seeing causation.

Composing Spatial Worlds

I   *So, in this focus on creating spatial worlds for the listener, are you seeking to communicate through their awareness and understanding of spaces and their relationship to and position within those spaces?*

NB   This is why I am working with ambisonics, because I want to put the audience into the spatial perspective of the music, for them to be part of that spatial relationship within the work and the materials.

In an ideal listening environment, the audience experience a physical connection with the projected space. This interest came from my experiences of the Loudspeaker Orchestra sound diffusion, so it is not only tied to the technology of ambisonics. I absolutely resonate with Jonty [Harrison]'s suggestion that diffusion performance is the 'life stage' of the composition.[151] When composing in stereo, you feel and understand what the special perspectives are; in the studio, the perfect phantom image[152] reveals a sense of 3D. In performance, you can unfold the spatial perspective over the Loudspeaker Orchestra. I have extended some of the diffusion performance principles into my work with ambisonics.

I   *Could we talk about questions of reality and abstraction articulating sounds in or-der to communicate emotion and communicate with people, and how it relates to these?*

NB   There is an underlying topic here, which is that my goal is to create music. I am not wanting to make things that are 'matter of fact' or a presentation of assembled sound recordings. To me, music comes from emotional expression.

Dirty Sounds and Emotion

I   *One of the things that other people have mentioned is a notion of emotional authen-ticity that comes through sounds that have a noisier character. Is that something that resonates with you?*

NB   Our emotions and experiences, all the things that we do, could be described as the dirtiness of life. Nothing ever follows a perfect description of how things should be. I feel like the dirtiness of life is very much a metaphor for the actual experience of life. Music is, for me, about counterpoint, about composing, and about putting things together. And when you put things together, you show their differences, or their contrasts, to avoid uniformity or a sense of grey. Recording a sound with clarity will capture its dirtiness.

Recording and Performance

I   *Do you think of recording in a performative way?*

NB   Yes, indeed. Being performative often means moving either the microphone or the source. But as soon as you start moving the microphone, there is technical noise to be aware of. I don't want you to hear my hands on the microphone, or anything else artificially interrupting the process. So, in the studio, I often move the source. Outdoors, that's more tricky! When I am recording environmental sounds, I would describe it as more exploratory. Often the microphone is stationary and the land-scape is the performer. But sometimes, I need to move the microphone or my com-plete body. An important difference between performativity indoors and outdoors is the time span. Outdoors involves greater time spans: the need to be patient and to let the landscape unfold at its own pace (which is something I have been exploring through a recent project *"Reconfiguring the Landscape"*[153]).

I   *I think this comes back to what you were saying earlier about this appreciation of us being one element within a wider system, rather than a totally anthropocentric focus on the human as being the centre of everything, it is about existing within a wider system and environment?*

NB   Yes. I think there is enough ego without making it worse [laughter].

Notes

1 We are grateful to Pro Sound Effects in providing access to these libraries as part of our research – https://shop.prosoundeffects.com/collections/ann-kroeber-alan-splet.

2 Nagra is a brand of high-quality portable audio recording devices that captured sounds on reel-to-reel 1/4" magnetic tape.

3 First experience of field recording while working at the United Nations.

4 Flat Response Audio Pickup – a form of triaxial contact microphone built by Arnie Lazarus – FRAP. See Lazarus (1974) for full details: https://www.aes.org/e-lib/browse.cfm?elib=2539.

5 *Dune*. (1984). Directed by David Lynch [Feature film]. US: Universal Pictures.

6 16 mm film (magnetic stock).

7 *Game of Thrones*. (2011–2019). Directed by David Benioff et al. [TV Series]. United States: HBO.

8 'Walk of Punishment' (2013). *Game of Thrones,* Season 3, Episode 3. Directed by David Benioff. [TV Series]. HBO, April 14, 2013.

9 Positive learning experiences, e.g., in life the greatest moments of learning have been the result of…

10 *The Lord of the Rings: The Rings of Power*. (2022). Directed by Payne, J.D., and McKay, P [TV Series]. California, US: Amazon Studios.

11 *Lovecraft Country*. (2020). [TV Series]. New York, US: HBO.

12 Alejandro Iñárritu – Film Director. Profile on IMDB: https://www.imdb.com/name/nm0327944/.

13 *The Conversation*. (1974). Directed by Hackman, G., Coppola, F. F., Cazale, J., Forrest, F., Williams, C., Roos, F., & Shire, D [Film]. Hollywood, California: Paramount Pictures Corp.

14 *Once Upon a Time in America*. (1984). Directed by Sergio Leone [Film]. United States: Warner Bros and Italy: Titanus.

15 A pattern of motif which persistently repeats throughout a passage.

16 *Once Upon a Time in the West*. (1968). Directed by Sergio Leone. [Film]. United States: Paramount Pictures and Italy: Euro International Films.

17 Sergio Leone (1929–1989), an Italian Film director, Producer, and Screenwriter, widely recognised as one of the most influential directors in the history of film.

18 Ennio Morricone (1928–2020), an Italian Composer, Orchestrator, Conductor, and Musician, with more than 400 scores for television and cinema composed, and considered as one of the greatest Film Composers of all time.

19 Ren Klyce (b.1987) is a Japanese-American Sound Designer and Sound Mixer. Online IMDB profile: https://www.imdb.com/name/nm0460274/.

20 David Andrew Leo Fincher (b.1962) is a Director based in the United States. His notable films include *Seven* (1995), *Fight Club* (1999), and *Panic Room* (2002).

21 John Milton Cage Jr. (1912–1992) is an avant-garde composer of indeterminacy in music, electroacoustic music, and non-standard use of musical instruments.

22 *Bohemian Rhapsody*. (2018). Directed by Bryan Singer. [Film] US: 20th Century Fox.

23 In sync.

24 Volume unit meter is a device that displays a representation of the signal level in audio equipment.

25 All the extras are always miming on set to protect the voice of the principal actors.

26 Vocal Groups, also referred to as Loop Groups, are collections of actors brought in to act out and record crowd sequences.

27 The COVID-19 pandemic was a global health disaster that necessitated the introduction of wide-ranging social distancing protocols to contain the spread of the disease. Individuals had to maintain distance from one another, which presented unique challenges for the recording of groups.

28 Queen Official. (2018). *Put Me In Bohemian!*. Upload 29 June. Available at: https://www.youtube.com/watch?v=aCDUSjR0uf0 (Accessed: 3 July 2022).

29 Queen. (1975). Bohemian Rhapsody. *A Night at the Opera* [CD]. United Kingdom: EMI.

30 *Fantasia*. (1940). Directed by Armstrong, S., Algar, J., Roberts, B., Satterfield, P., Sharpsteen, B., Hand, D., Luske, H., Handley, J., Beebe, F., Ferguson, N., Jackson, W. [Film]. California, US: Walt Disney Productions.

31 *Gravity*. (2013). Directed by Alfonso Cuarón [Film]. California, US: Warner Bros. Pictures

32 *Everest*. (2015). Directed by Baltasar Kormákur [Film]. US: Universal Pictures.

33 *Mowgli: Legend of the Jungle*. (2018). Directed by Andy Serkis [Film]. California, US: Netflix, Inc.

34 An English colloquial term meaning upset.

35 The cumulative non-human sound produced by living organisms in a large area characterised by its soil, vegetation, climate, and wildlife.

36 A naturally occurring sound produced by a habitat produced by non-living elements, e.g., the wind and the waves.

37 Vlad, G. (2021b). *The Weird Sound of Empress cicadas*. Available at: https://youtu.be/6ZPDYlvj7MM (Accessed: 1 August, 2022).

38 Link to George Vlad's Bandcamp page: https://bandcamp.com/georgevlad.

39 Link to George Vlad's YouTube page: https://www.youtube.com/c/georgevlad.

40 Drop rig is where recordings are captured in the field without the recordist present. The equipment is set up and left behind to capture sound that may occur over time.

41 A ravine is a deep valley or gorge that is often narrow with steep sides.

42 A tropical grassland with warm temperatures all year round and seasonal rainfall.

43 An effect in which when sound reflects off the curved surface, it will bounce in a straight line no matter where it originally hits.

44 A short single-repeat echo effect.

45 Vlad, G. (2018). *Thunderstorm At Langoue Bai*. Available at: https://soundcloud.com/georgevlad/thunderstorm-at-langoue-bai?in=georgevlad/sets/congo-basin-rainforest-2018 (Accessed: 1 August, 2021).

46 HRTF = Head Related Transfer Function. A calculation that approximates the delay and filtering of the human head to reproduce an impression of binaural listening.

47 *Room*. (2015). Directed by Lenny Abrahamson [Film]. Canada: Elevation Pictures, UK and Ireland: StudioCanal, United States: A24.

48 Caoimhe Doyle is a Foley artist based in Ireland - https://www.imdb.com/name/nm0236304/.

49 *Normal People*. (2020). Directed by Lenny Abrahamson & Hettie Macdonald [TV series]. UK: BBC Studios.

50 A form of contact microphone under the floor to capture low-frequency components of steps and hits.

51 Boom is a shorthand description for the directional microphone suspended on the end of a boom pole, used by the 1st Assistant Sound (Boom Operator) to capture dialogue when filming scenes. Lav is a shorthand description of a lavalier microphone, which is a small microphone attached to, and often concealed, on an actor's person to capture up-close recordings of dialogue.

52 "According to urban legend, I am "the first and only" black female "Post Production Sound Department ADR/Foley Re-Recording Mixer" working at this level on the planet." – https://cinemontage.org/cut-to-black-jesse-dodd-adr-and-foley-re-recording-mixer/.

53 Richard Anthony Wolf (b.1946) is an American Television Producer.

54 *Law and Order*. (1990–2010, 2022 – present) [TV series] US: National Broadcasting Company.

55 *One Chicago*. (2012–2022) [TV Series] US: National Broadcasting Company.

56 Founded in Manhattan in the 1950s, the electronics manufacturer specialised in magnetic film and sound recorders, Post-Production cinema film equipment, and 16mm and 35mm film projectors.

57 A combination of software, hardware, and network protocols that delivers uncompressed digital audio over an Ethernet network.

58 A large diaphragm condenser microphone.

59 A type of shotgun microphone.

60 A type of shotgun microphone.

61 The sounds of a given location or space.

62 *Chicago PD*. (2014 – present). [TV series] US: NBC Universal Syndication Studios.

63 *The Underground Railroad. (2021). Directed by Barry Jenkins [Drama Series]. US: Amazon Prime Video.*

64 Dolby Atmos is a surround sound technology developed by Dolby Laboratories.

65 Nicholas Britell (b.1980) is a Composer, Pianist, and Film Producer based in New York City.

66 Barry Jenkins (b.1979) is an American Filmmaker.

67 *Game of Thrones*. (2011–2019). [TV Series]. US: Warner Bros. Television Distribution.

68 *Barton Fink*. (1991). Directed by Joel Coen [Film]. US: 20th Century Fox.

69 Amy Beth Schumer (b.1981) is an American stand-up Comedian and Actress.

70 Daniel Robert Elfman (b.1953) is an American Film Composer, Singer, and Songwriter.

71 *Mufasa: The Lion King*. (2024). Directed by Barry Jenkins. [Film]. US, Disney +.

72 *The Killing of Two Lovers*. (2021). Directed by Robert Machoian [Film]. US: Neon.

73 Academy of Motion Picture Arts and Sciences.

74 Sir Alfred Joseph Hitchcock KBE (1899 – 1980) was an English Filmmaker, regarded as an influential figure in the history of cinema.

75 The raw unedited footage shot during that day.

76 Joel Daniel Coen (b.1954) and Ethan Jesse Coen (b.1957) are American Filmmakers.

77 *The World Soundscape Project* (WSP) was established as a research group by R. Murray Schafer at Simon Fraser University during the late 1960s and early 1970s. For more information visit: https://www.sfu.ca/sonic-studio-webdav/WSP/index.html.

78 Schafer, R. M. (1994). *The Soundscape: Our Sonic Environment and the Tuning of the World.*

79 Westerkamp, H. (1975). Whisper Study. *SFU 40* [CD]. Burnaby, Canada: Simon Fraser University.

80 Founded in 1993, *The World Forum for Acoustic Ecology* is an international association of organisations and individuals whom share a concern for soundscapes across the world.

81 Westerkamp, H. (1999). Moments of Laughter. *Breaking News* [CD]. Canada: earsay music.

82 Wolfgang Amadeus Mozart (1756 – 1791) was an influential and prolific composer of the Classical Period in Music.

83 Westerkamp, H. (1992a). A Walk Through the City. *Transformations* [CD]. Canada: Empreintes DIGITALes.

84 A weekly programme of Contemporary New Music from 1978 to 2007 on CBC Radio.

85 Westerkamp, H. (1989). Kits Beach Soundwalk. *Transformations* [CD]. Canada: Empreintes DIGITALes.

86 Westerkamp, H. (1993–2000). *From the India Sound Journal.* Available at: https://www.hildegardwesterkamp.ca/sound/comp/2/indiasj/ (Accessed: 1 September 2021).

87 Westerkamp, H. (1987). Cricket Voice. *Transformations* [CD]. Canada: Empreintes DIGITALes.

88 Pierre Henry (1927–2017) was a French Composer and pioneer of Musique Concrète.

89 Vande Gorne, A. (1984). Eau. *Tao* [CD]. Canada: Empreintes DIGITALes.

90 François Bayle (b.1932) is a composer of Electroacoustic Music and coined the term Acousmatic Music.

91 Vande Gorne, A. *Treatise on writing acousmatic music on fixed media.* Ohain, LIEN vol. VIII, 2017 80p. (first French edition), and vol. IX, 2018 (English translation) p.14–23.

92 François Delalande (b.1941) is a theorist of Electroacoustic Music. He was Director of Research in Music Sciences for 30 years at the GRM, Groupe de Recherches Musicales in Paris.

93 Treatise on acousmatic composition, in progress, LIEN vol XII to be published in 2024.

94 Hans Tutschku (b.1966) is a Composer of Instrumental and Electroacoustic Music.

95 GRM Tools Spaces is a suite of plug-ins that allows for exploration and fragmentation in the use of space.

96 Benjamin Thigpen (b.1959) is a Composer and Sound Artist, composing works in the genres of Noise and Electroacoustic Music.

97 Lemur is a controller app that allows the user to control anything, e.g., electronic music performance software, DJ software, and stage lighting.

98 Studer is a designer and manufacturer of professional audio equipment used for broadcasting and recording studios.

99 Musique Concrète is a form of music composition that uses recorded sounds as the raw material.

100 Pierre Schaeffer (1910–1995) was a French Composer, Acoustician and Electronics Engineer. In 1948 with the Radio-diffusion et Télévision Française, introduced the Musique Concrète.

101 Igor Fyodorovich Stravinsky (1882–1971), an influential composer of the 20th century.

102 A visual programming language for the creation of music and multimedia works.

103 Burt T, W. (1996). Some Parentheses Around Algorithmic Composition. *Organised Sound*. Cambridge University Press, 1(3), pp. 167–172. https://doi.org/10.1017/S1355771896000234.

104 Gérard Grisey (1946–1998), a French Composer at the centre of 'Spectralist' thinking. Masterpieces include his *Partiels* (1975) and *Vortex temporum* (1994–96).

105 John Cousins (b.1943) is a Composer based in Christchurch, New Zealand.

106 Denis Arthur Smalley (b.1946) is a Composer of Electroacoustic Music (with a special interest in Acousmatic Music).

107 Young, J. (2007). Ricordiamo Forlì (2005). *Lieu-temps* [DVD-A]. Montréal: Empreintes DIGITALes.

108 Young, J. (2010). Lamentations. *Metamorphose 2010* [CD], Ohain: Musiques & Recherches.

109 *Young, J. (2020). Once He Was a Gunner (2020). Histoires des soldats [CD]. Montréal: Empreintes DIGITALes.*

110 The earliest known device designed to record sound. Invented by Edouard Leon Scott de Martinville in Paris in the 1957.
111 Wilkinson, A. (2014). *A Voice from the Past*. Available at: https://www.newyorker.com/magazine/2014/05/19/a-voice-from-the-past. (Accessed: 30 November 2021).
112 Iannis Xenakis (1922–2001) was a Romanian-born Greek-French Composer and Music Theorist of Avant-garde Music.
113 John Christopher Williams AO OBE (b.1941) is an Australian virtuosic classical guitarist.
114 An orchestra of loudspeakers arranged in front, around, and within the audience in a concert hall setting. It was designed and inaugurated by François Bayle in 1974 and is often used for the presentation of Acousmatic works.
115 Wishart, T. (1996). *On Sonic Art*. 2nd edition (Edited by Simon Emmerson). London: Routledge.
116 Wishart, T. (1994a). *Audible Design: A Plain and Easy Introduction to Practical Sound Composition. York: Orpheus the Pantomime Ltd.*
117 Composers Desktop Project - https://www.composersdesktop.com.
118 Playing technique that involves plucking the strings of a stringed instrument.
119 Playing technique that involves bowing the strings of a stringed instrument.
120 An extended technique in which several notes are produced simultaneously on a monophonic musical instrument.
121 Wishart, T. (1994b). Tongues of Fire. *Tongues of Fire* [CD]. UK: Orpheus The Pantomime.
122 Wishart, T. (2014a). Globalalia. *Globalalia / Imago* [CD]. UK: Orpheus The Pantomime.
123 Wishart, T. (2014b). Encounters in the Republic of Heaven. *30 Jahre Inventionen VII 1982–2012* [CD]. Germany: Edition RZ.
124 Choral works, settings of the ancient Atra-hasis story of the flood myth, composed as part of the *Over Lunan* project – available to listen here: https://www.ahillav.co.uk/atra-hasis-2021/
125 Ladefoged, P., & Maddieson, I. (1995). *The Sounds of the World's Languages - Phonological Theory*. Hoboken, NJ: Wiley.
126 Axiomatic = self-evident, obvious, accepted.
127 A person who solicits business or employment in an annoying and persistent manner.
128 Matatus are privately owned minibuses used as shared taxis. They often use decoration and music to attract customers.
129 John Cage is interviewed as part of the film *Listen* (1992) by Miroslav Sebestik.
130 A group of three jazz musicians.
131 Details of KMRU's compositions can be found online: https://kmru.info/.
132 A library of pre-recorded sound effects with the intention of them being reused for different purposes.
133 A type of microhpone that senses audio vibrations through contact with solid objects.
134 Ambisonic microphones capture 3D sound fields using an array of microphone capsules.
135 Mahtani, A. (2019a). Aeolian. *Racines* [CD]. Canada: Empreintes DIGITALes.
136 Mahtani, A. (2019b). Past Links. *Racines* [CD]. Canada: Empreintes DIGITALes.
137 Recent dance scores:
  And Breathe. (2021). Directed by Sima Gonsai [Dance Film]. UK: Sima Gonsai Films.
  Romeo and Juliet. (2021). Directed by Rosie Kay Dance Company [Dance]. UK: Rosie Kay Dance Company.
  10 Soldiers. (2019). Directed by Rosie Kay Dance Company [Dance]. UK: Rosie Kay Dance Company.
  MK Ultra, Rosie Kay Dance Company, 2018
138 *Tattoo Stories. (2021). Directed by Steve Johnstone. [Podcast]. Smethwick: Black Country Touring. https://bctouring.co.uk/tattoo-stories-podcast/.*
139 *Baby Daddy. (2018). Directed by Elinor Coleman [Theatre]. UK: Birmingham Repertory Theatre.*
140 Modernist art movements from which Electroacoustic Music sprang often demand a conceptual rationale to validate artistic work. Students are often made to create a conceptual idea from which their work is supposed to extend.
141 Emmerson, S. (2001b). *Pentes: A conversation with Denis Smalley. SAN Diffusion*. Available at: https://electrocd.com/en/presse/428_5 (Accessed: 1 September 2022).
142 A process in which an audio sample is broken down into tiny segments of audio.
143 Denis Smalley references Edward Hall's four zones of perception space – public, social, personal, and intimate (Smalley 2007).

144   Stavropoulos, N. (2018). *Inside the Intimate Zone: The Case of Aural Micro-Space in Multichannel Compositional Practice.* In: Proceedings of the 15th Sound and Music Computing Conference (SMC2018). Sound and Music Computing Network, pp. 113–117. ISBN 978–9963-697-30-4.

145   Stavropoulos, N. (2016). *Topophilia.* Available at: https://soundcloud.com/nikos-stavropoulos/topophilia (Accessed: 3 November 2022).

146   Stavropoulos, N. (2017). *Karst Grotto.* Available at: https://soundcloud.com/nikos-stavropoulos/karst-grottoexcerpt-stereo-reduction (Accessed: 3 November 2022).

147   Stavropoulos, N. (2019). *Claustro.* Available at: https://soundcloud.com/nikos-stavropoulos/claustro-stereo-reduction (Accessed: 3 November 2022).

148   Autonomous sensory meridian response is an emotional state triggered by whispering, light touch, etc.

149   Denis Smalley is a significant theoretician and Composer of Electroacoustic Music. His concept of Spectromorphology is a "collection of tools for describing sound shapes, structures, and relationships, and for thinking about certain semiotic aspects" (Smalley 2010, p.95). Spatiomorphology extends spectromorphology to consider also the spatial aspects of sound. It is a term that acknowledges the significance of space within electroacoustic music and provides a frame to enable its discussion and study (Smalley 2007, p.53).

150   Michel Redolfi (b.1951) is the Director of the Centre International De Recherche Musicale (CIRM) in Nice since 1986 and the co-founder of the Groupe de musique expérimentale de Marseille (GMEM) in 1968.

151   Harrison, J. (1999). *Diffusion: Theories and Practices, with Particular Reference to the BEAST System.* Montréal: Communauté électroacoustique canadienne / Canadian Electroacoustic Community. eContact!, 2.4 Multi-Channel Diffusion.

152   A phantom image can be created between two (or more) loudspeakers, creating the illusion of an additional speaker/s.

153   Link to further information on the project: https://www.researchcatalogue.net/view/1724428/1724429.

5 Dialogues

In the dialogues that follow, you will find co-edited conversations developed in partnership with each contributor, with verbatim texts of our discussions edited for clarity and factual accuracy. Specialist terms are explained within footnotes so as not to disrupt the flow of the original conversations.

Each interview has been segmented with subheadings to categorise key topics and highlight significant thematic ideas.

List of Dialogues

DOI: 10.4324/9781003163077-5

5.1 Ann Kroeber with Hildegard Westerkamp

Ann and Hildegard are two female pioneers of sound practice with a shared sensitivity to listening to the environment and the communicative potential in animals and insects. Their work captures recordings that are impressions from the natural world, rendering powerful new perspectives that transform our perceptions and enable us to hear in new ways.

Listening Excerpts Shared

To stimulate the conversations, we invited each participant to share up to ten minutes worth of examples that exemplify their practice.

Ann Kroeber:
   Selected recordings from the Splet/Kroeber sound library[1]

Hildegard Westerkamp:
   A Walk Through the City[2] (extract) – 3 minutes, 24 seconds
   Attending to Sacred Matters[3] (extract) – 5 minutes, 34 seconds

Speaker Key:

I   Interviewer
AK Ann Kroeber
HW Hildegard Westerkamp

AK   Hildegard, I was simply stunned by your soundscapes. I couldn't find the words to tell you how imaginative and beautiful they are. Thanks so much for sharing them with me. I particularly enjoyed, *A Walk Through the City.*

HW   I had a similar reaction to your recordings. Thinking "oh my goodness, they are so clear and so beautiful".

AK   Thank you.

HW   What I find interesting is that we have something in common in terms of beginnings. I read some previous interviews[4] that you gave in which you mentioned that your discovery of a passion and talent for field recording just happened to you. There wasn't necessarily a plan or conscious choice and that's precisely what happened to me as well. I came to soundscape composition through a focus on listening to the environment.

I   *Ann, I wonder if you feel that Hildegard's piece "A Walk Through the City", which is exploring the urban environment and capturing the sounds in the space, is in some ways similar to the story that you tell about your experiences of recording fireworks walking through the city on Chinese New Year?*

AK   Yes, that's an interesting observation. I think maybe so. Because the sounds that Hildegard captures are really quite unique in a way and there is something special about them. Was there something about the sounds you chose that appealed to you Hildegard?

HW   Once I started field recording, I discovered that a whole other thing happens to your listening, your ears wake up in a new way. It unfolded in stages, the first was coming out of music studies and being introduced to a lecture from [R.] Murray Schafer, in which he proposed that one should listen to the whole environment as if it was music.[5] That woke up my ears and then field recording became a natural follow-up.

I   *Did you feel, Ann, that you had a process of discovery like that? Are there any moments that you remember, in your career, where you really discovered something new or a new perspective, or a new experience that changed how you listened or worked?*

AK   Every time I went out recording, I had that experience [laughter]!.

HW   No, I totally agree. And I like what you say about "every time you record it is a new experience". Because when you're recording environmental sounds, you are constantly experiencing a totally new situation. A totally new context.

AK   That's right. Yeah.

HW   So, there is that first *aha* moment of "oh, I really like this", which is the discovery that starts it all off. And that process of discovery happens over-and-over again in your life, because you are so connected to the process of listening.

AK   Yes. Absolutely.

HW   When I read your interview, I thought, "Oh, I need to talk to this woman, she has the same experiences I have" [laughter].

Sound Recording – The Process

HW   I'm very curious about how you record because your context is so completely different than mine. You are recording for sound design and film, and your ear leads you from that perspective, and so I assume that you're aiming to get the clearest possible isolated 'sound effect' result that gives you a certain mood for the film that you're working on. Is that correct?

AK   Well, when I record, I don't think about a lot of specifics. I'm just trying to be there and listen, and things that move me, I record.

HW   Okay, so we have that in common actually, it's the same for me when I record, the end result is not really part of my thinking process. But it's interesting for me to hear that it's the same for you, because the recordings that you sent are all very specifically about one object, for example, an animal or a machine. There's not necessarily a wider context audible, well mostly… The impression I got is that it was more focused on a clarity in foreground sound.

AK   Well, when I'm recording horses, I don't turn my mic around and go "oh, that sounds good" and listen to some wind off in the distance. I stay very focused on horses, but in doing that, I talk to them and let them express themselves.

HW   Yes, I enjoyed reading that in your interview from 2011, on how you connect with them and make your recording accordingly, so that the atmosphere of that situation is good for everyone.

Actually, you mentioned the horses, the quality of those recordings is fabulous.

AK   My husband and I recorded those, we put microphones under the horse's belly and covered them with Styrofoam. We wanted to capture the sounds of galloping from the horse's belly, getting that perspective very close to the horse's hooves. And that really has worked well. So, it's different techniques for different purposes.

Communicating with Directors

HW   I have worked in film only very, very little. And I found it difficult at times. I'm not sure whether it takes a special skill to adjust to that kind of collaboration in a film, where you have to speak about sound with a Director, and you have to communicate in ways that you both understand.

AK   I was very fortunate that most of the Directors that I worked for were sensitive to sound, and were open to my suggestions. I would do creative edits and then surprise them and let them hear it and find out what they thought. Sometimes, it gave ideas for doing new things in the film, especially with David Lynch[6] who might say "oh, we could take that sound and use it for 'this' or 'that'".

Sometimes, the sounds I had gone out to record may not work. So, we would end up using other material that I've recorded previously, and is in my sound library. That too might inspire new ideas.

HW   In my practice, I find it difficult to take sounds out of their original contexts and transpose them into the new context of the film, as atmosphere for film.

I feel like it's okay for me to do this in my compositions, because there's a continuity between the experience of recording and being in the studio and working with those sounds for a composition. But it is more difficult to do the same for film.

AK   Martin Stig Andersen[7] contacted me aeons ago when he was working on *Wolfenstein II: The New Colossus*.[8] He was looking for stuff like machines, foghorns, and animal sounds that he could incorporate into his compositions. I gave him a bunch of things and I told him how he might like to use them in his project. He was so excited, and in the end, he earned several award nominations for it. It was great. It was really sweet. And he was one of the few people that credited me.

So, you can never know what your sounds might be used or useful for. You can never tell!

HW   I suppose you see a scene, or you see a film, if you have a library of sounds to draw from it doesn't need to belong to the context in which you recorded it, it can go anywhere.

AK   Yes, absolutely.

HW   I want to to tell you about an experience I had with Gus Van Sant,[9] who was interested in excerpts from one of my compositions. So, I did not do any sound design for him; he just wanted to use a piece that I had already composed, for the film *Elephant*,[10] about the Columbine High School disaster. To my mind, my piece *Beneath the Forest Floor*[11] was very much about peace, peacefulness, and stillness. And I was concerned about my work being used in violent scenes. But I was assured by his Producer that I could rely on Gus's judgement.

When I saw the film, I discovered that the excerpts that he used from *Beneath the Forest Floor* were placed precisely where the first shooting starts! And to my great surprise, it worked beautifully. And I was trying to figure out why. In doing so, I realised something about my own piece, of which I had not really been conscious of that there's a lot of darkness in it. It really taught me a lot, because I was convinced that I had connected with the entire context of this place when I recorded and created a piece about it. Yet, Gus van Sant took this piece out of that original context and put it into his film, because with his filmmaker's ears, he heard the darkness that he was searching for, and it was absolutely incredible how well it worked. The Film Director's ears were incredibly astute and had listened to my piece from a completely different perspective.

AK   Yeah, that's cool. I've had that experience a lot.

Sensitivity to Sounds

I *I wonder whether I can come back to this question of sensitivity to sounds, and the types of sounds that you are both drawn to capture. Hildegard, when we spoke, you told me about the piece "Moments of Laughter",*[12] *using the sounds of your daughter's voice and about some of the reactions to the emotional intensity of that piece. And I feel that lots of Ann's recordings also have this very strong emotional intensity.*

I wonder if within both of your practices there is a deep relationship with the other?

HW Like intimacy?

I *But also perhaps about trying to create the other perspective, or share the perspective of the other?*

AK Yes, absolutely. Get out of the way, get your ego out of the way, and just be there with them, and realise that there's so much more going on in terms of communication. It's not just a sound effect, there is so much interaction and the communication is beautiful. But you have to keep yourself out of it, stop being all full of yourself. It's not just about recording sounds, it is absolutely a two-way thing.

HW And when that space is created, then animals often do come to you.

AK They do.

HW I had one experience in the desert in Mexico, where we were in a place called the *Zone of Silence*,[13] a bunch of artists gathered there, and I was the only Composer, and I was recording, and the desert was very quiet and there weren't many sounds to record in the day. But the nighttime, of course, was the interesting part with the crickets. And so I went away from the camp into the desert, and just sort of sat down in the middle of crickets in the dark and, you know, the sky was, the stars, and it was just beautiful and there was no other sound, it was just the crickets. And at one point, and I had probably been out there, maybe for ten minutes, and suddenly, it was almost as if – this visual image I have – this cricket steps forward out of the chorus, it comes right up to my microphone and starts making its sound and that lasted for two minutes. And then, it moved away a bit and I tried to, I couldn't see it, I tried to find it again. And I found it for a short time and then it was gone. But it's a solo recording of a cricket. And that cricket, of course, became the main character in my piece *Cricket Voice*.[14] But that was a magical moment. I just literally froze when this happened and didn't dare to move. It lasted for about two minutes, and it was very magical.

Reciprocity Between the Subject and Recording

AK Many years ago, I was in a zoo, and I was recording different creatures, one of which was monkeys. There were these medium-sized monkeys, a little bit smaller than chimps. They lived on a really high rock called "Monkey Island". And I came down to their enclosure and I said, "hey guys, come down, I want you to hear something", and the whole troop came down single file from the rock and they sat around me like kindergarten kids, with crossed legs as if to say, "yeah, what do you want to do?".

I had my Nagra,[15] which meant I was able to play back sounds I had recorded earlier. And I played these chimps a recording of some smaller Dusty Titi monkey's that I had made in a different part of the zoo. They made these little sounds like "teep, teep, teep, teep, teep, teep, teep, teep, teep, teep". Anyway, I played it for the chimps, and they listened really closely. After the recording I said, "I bet you guys can do better". And the whole troop jumps up as if to say "yeah we can" and they started hooting and hollering around me making a racket. It was so great, I just wanted to cry. I was so touched by them, one actually got on my shoulder and hoots into the microphone. That was such a great recording.

HW   That sounds amazing.

I   *I wonder whether there's something interesting to explore there in terms of the character of the sounds and how these recordings might not be used in a documentary fashion for just being monkeys, they might be used for being something else? But there's something about that liveness, and that emotion, and that energy within the recording that you're capturing and transferring.*

AK   Gosh, I don't know. That's a little too intellectual of a question to ask me [laughter]. I'm just there in the moment. These monkeys fascinated me, with the way in which they communicated. I was listening to what they had to say and how they responded and who they were. It was just beautiful. And I'm always touched by how smart animals are and how expressive they can be.

HW   Do you remember where those recordings have been used?

AK   I don't remember exactly. They have been used for various things. You can take sounds and manipulate them so much. So, I honestly don't remember the specifics. But I record a lot of other things besides animals, for example, I recorded a lot of industrial sounds for David [Lynch]. With *Dune*,[16] I applied the same approach that I might use recording animals when I went around and listened to industrial sounds. All these, clanks and bangs and rhythmic noises they transported you into a different space.

HW   It's interesting because when I think about my cricket recording, I wanted to honour that moment of magic and retain a sense of the original experience. How do you work with such a sound to keep, or even open up, that magic?

AK   Working in film, I had to let that go. I have the memory of it, certainly forever, but it becomes something different when you have recorded it, and I don't hold on to it so much that I won't allow you to do something else with it. I try not to be too precious.

HW   In that context, you can't hold on to it, because it's going to go somewhere else, right?

AK   The sounds can change and be used for something completely different.

HW   I think that's maybe quite a crucial point of difference here. When you work for film, you have to be able to let go, and I had a really hard time with that.

AK   Absolutely, yes. And I don't feel like I'm being disrespectful, but maybe I should? Maybe they wouldn't have been happy if they knew they were going to be used in something totally different [laughter].

I   *Do you think that you help people to listen more through the work that you do?*

AK   I don't know. I mean, I hope so. People don't pay attention to sound, they're not conscious of it, and so they just dismiss it. They don't realise how much of an effect the sound has on their feelings in a particular movie. I mean, it just makes all of the difference in the world.

    It's so frustrating when someone just disregards what everyone did and they perhaps call it noise, or they refer to everything as Foley, all of this recording that's going on in the real world and they think it's Foley!!! Come on, please. It just drives me crazy!

HW   Yeah. And it has to do with the state of listening in general, the visual has overpowered, and we need to get in touch with our listening. My experience has been when I've made students aware of the soundtrack they get alot more out of their film experiences. Rather than unconscious connections with sound, they make conscious associations, and that's very enriching.

5.2 Nina Hartstone with Trevor Wishart

Nina and Trevor are leading experts/practitioners on the expressive potential of the voice, and its articulations within and beyond language as a powerful communicative tool. Though operating in different contexts, we wanted to bring together two practitioners who each had such a rich understanding and deep sensitivity for articulating creative communication via the power of the voice.

Listening Excerpts Shared

To stimulate the conversations, we invited each participant to share up to ten minutes worth of examples that exemplify their practice.

Trevor Wishart:
  8 minute 41 seconds of extracts from:
  Preacher, from *Two Women*[17]
  "I'll always love dance" from *Encounters in the Republic of Heaven*[18]
  Opening and extract from *The Division of Labour*[19]
  Fourth "Negotiations" section of *The Garden of Earthly Delights*[20]

Nina Hartstone:
  Excerpt from *Mowgli: Legend of the Jungle*[21] – Timecode: 00:06:10–00:09:50
  Excerpt from *Gravity*[22] – "Detached" Scene – Timecode: 00:11:54–00:17:16

Speaker Key

I   Interviewer
TW   Trevor Wishart
NH   Nina Hartstone

I   *Nina, I wondered if you could share any reactions from listening to Trevor's work. Were there elements that you connected with and recognised from your own practice and working with the voice?*

NH   It's interesting to hear the human voice used in different ways, whether it's to do with pitch, or rhythm, or shape. Given what I do, I'm always listening to the detail within every little syllable, and the feelings that vocal sounds can evoke.

It's just so interesting to think of the potential of the human voice. We're very used to using it for creatures or monsters, but to treat it as an element on its own seems quite similar to processes of sound design. How we might take the roar of a lion or a tiger and mix it underneath a car engine. Using the voice as a sound, rather than in relation to its source.

This said, there seemed to be a crossover at points, sometimes the qualities of individual sounds meant that they were somewhat disconnected from the source, while at other times I felt like the sound was very connected to the source, and it was being used for that reason.

There was a section in *Encounters in the Republic of Heaven* where it sounded like parts of the voice had been cut, with spaces added. When you hear these segments in fast succession you can almost hear what they were originally saying, but at other points the timings of the cuts shifted, and it sounded like the speakers were saying something completely different. I found those elements fascinating, as well as the transitions using syllable sounds like 'sssss's' to transport you to a different scale in a way that provides a different emotion, and different kind of response within you.

It was very interesting the emotional responses I had to the different pieces.

Compositional Process/Structure

TW That's great to hear! Personally, I think the only point of music is its emotional impact [laughter]. When I'm working, I'm very constrained by the formal notions that are guiding that particular project – for example, the notion might be, 'making a piece using syllables from an array of different languages' – but I am constantly trying to elaborate the emotional within these constraints.

I'm not doing so intellectually, analysing my reaction etc.. I'm listening to sounds and thinking about what is the emotional and physiological effect on me? I listen to (A) and it has the right structure and it grips me, where (B), has the right structure and it's boring. It comes in a natural, intuitive way. And if those things are not there, the piece doesn't work, it's just boring, formal and no one will want to listen to it.

What I hate is music which is very cleverly put together and it's just boring. You go to some contemporary music concerts and there is this is a long programme note about how they used this structure or used set theory and this computer programme, and yet you listen to it and it's boring because that is not enough.

There's something else intangible about music, which is the question of: Why do we listen to it? Why do we value it? How this grips you in some mysterious way which has, for me anyway, nothing to do with the story. It's a kind of narrative perhaps, but it's not a narrative.

NH It's so interesting because, in film we are playing so much with the context and the impact that shifts will have on you. Particularly with the score entering when you haven't had it for a time, it's sometimes more about the fact that it has appeared where it wasn't before.

You build up the audiences' expectations about what is going to come next, but if you flip that expectation on its head, you can change the context of what happened before. You can be very effective with sound in these moments, it often actually puts the viewer back into an active listening mode and directs them to pay attention to the next part of the story.

TW The way music *really* works is that you remember what happened before and you relate that to what is happening now. Musical structure is essentially playing with the memory of events – the similarities and the differences.

There are only two things you can do in music. You can either repeat something or you can change it. All music can be analysed in those terms [laughter].

It sounds very abstract when we talk about it, but it's not somehow. For some reason, it relates extremely strongly to our emotions. We can point to it working, but why it works nobody knows, and that's why I do music, because that's what interests me about music.

Music = Emotion

NH I certainly was feeling an emotional response to the samples that you sent through.

TW That's really good! If I think about your work Nina, I suppose the one thing that interested me the most in the clip from *Gravity* was this sense of cinematic space, or what you might call the acoustic landscape. I'm very interested in spatial things in Electroacoustic Music, and I was struck by the idea that the whole mix was very much a spatial illusion created for the listener.

It's not as if you're sticking your microphone out and this is what you get. In reality, astronauts would hear their own voice and the others in headphones in mono and nothing else. They might hear some interference, but certainly nothing would move. And it would actually be very boring [laughter].

So, the sound design is creating this kind of double illusion. Firstly, that you can hear anything at all; because if you were there you wouldn't hear anything. And secondly, the movement of the sounds, all that movement, going on in reality, would not be taking place acoustically, so all of that is created for the listeners.

It was fascinating for to me to sit back and actually listen because, like most other people I suppose, I didn't ever stop to think about sound when I'm watching a film. You go to the cinema, you enjoy the story, the fantastic images and yet, of course, the whole thing depends emotionally, and in almost every other way on the sound which we take for granted. So, it was very interesting to stop and listen to what was going on.

NH   It certainly was a tricky one, *Gravity*. Balancing how to sonically support the storytelling and attention, but also trying to be faithful to the science of the fact that you don't hear in space [laughter]. In the end, there was quite a heavy crossover between sound design and music. Some of the musical scores sounded a bit more like sound design, and a lot of the sound design was quite tonal.

But the way that we approached it was to try and anchor it to the scientific facts around how sound behaves in space. We tried to present everything as if it came through their body, so there was a lot done with contact mics, so all of their movement is not how it sounds from the outside of their suit, but is maybe how they would hear the movement from inside the spacesuit.

Of course, there's certain sections where things are blowing up, you just can't have it completely silent, you've got to... [laughter] ....try and pick a few moments to just be cinematic. The same as when someone pulls their sword out, you know it's always exactly ['shing' sound]... it doesn't matter what's coming out of, it's this total cheat but people will be disappointed if it doesn't sound this way.

I   *I wonder if we could explore some of the elements in that sequence of 'Gravity', about how you're articulating the pace of the breath as a kind of metronome to shift the mood and the tensions throughout the film. Which is a fascinating way of communicating nonverbally, which I guess links to some of the ideas that Trevor is exploring with taking words and syllables and re-articulating them.*

NH   *Gravity* was one of those jobs that taught me an awful lot about breathing and how it expresses the things that are going on inside your body. It's as much about when you hold your breath, as when you breathe. A long inhale to a hold is very effective, which is what happens in that sequence we shared from *Gravity*. She holds her breath, and you think "is she ever going to start breathing again?!" She keeps holding it, and it's that which sustains the tension.

It's very primal isn't it? You can hear her breathing constantly, yet when you hear something stops breathing, in an instant you feel fear for that being.

But there was so much subtlety in the types of breaths that needed to be there to make the sequence work, and quite a challenge to record those breath performances on their own in an ADR theatre, which is what happened in *Gravity*.

You might feel that you could just cut to inhale and exhale, but there's so much more detail that goes on, which is crucial for making something that feels believable, but which also serves the story, adding that build of tension and providing the release.

When you're sitting there with your cue to do "anxious breathing through this scene", what you often end up with are these forced quivery breaths. They don't really reflect the true kinds of breath that you'd make when performing physical actions, and it felt like quite a tightrope to try and get exactly the right timbre of breath that worked visually for each moment, but also worked for the whole arc of the story.

Slow breathing is really the hardest because everyone ends up just sounding like they're asleep! [laughter].

Voice as Instrument

TW I feel that my personal interaction with those kind of breath gestures is through per-forming. I do free improvised performing with the voice and I use various strategies, some are to do simply with sound making, but sometimes it is to do with theatre. For example, when you breathe in, and you don't breathe out for a long time [holds breath in pause].

Extreme types of vocal gesture are perhaps even almost universal, if you [gasp] anywhere in the world everybody knows what's going on, so I can certainly see the art in that.

NH It's interesting what you say about how accessible it is and how everyone follows along, because one of the things that I noticed when working on *Gravity* is that I was constantly hyperventilating because I'm breathing with her! [laughter].

TW Yes [laughter].

I *Could you talk a bit about the decision-making process, how you know when you have got to where you want to be as a result of having an emotional reaction and response to the sounds?*

TW [Laughter]. I think people don't talk about it because [laughter] you can't pin it down. If we knew the answer to that that it would be great, but we don't!

NH [Laughter] It is almost impossible to understand why these things happen, isn't it?

There's something incredibly raw that goes beyond any form of description, which is human to human. I don't know whether it's to do with past memories that trigger something in you, to remind you of something? But it is clearly felt and I do use my own reaction to things to inform my work. If it's supposed to be an emotional scene and something makes me feel, I will definitely use those performances rather than anything else, even if another may be technically slightly better. My preference is always to think, "if this is this is doing what we need it to do in this scene, let's go with that."

Structure and Context

TW It relates to the materials and the context within the piece. Sometimes you listen and listen and think, "I know, that is right. I can't tell you why, but it is". And I guess the proof of the pudding is that other people think it's right as well…

There are these notions in music theory that somehow there are certain propor-tions which are perfect, and if you use those, you'll get the right answer. But my experience is that that is simply not true.

It's a nice thing to think and it looks great on paper – you're writing a score and you see that's 75 bars long, that's 43 bars long and that's a perfect relationship – but that's not what music is. Music is what you hear. And so if you're working acoustically, then you become very aware of having to make these judgments about what works well, but I can't tell you anything about it. Just that I do it all of the time [laughter].

One of the big experiences I have is you can't make the whole piece at once, you make it in bits and those bits sound really great, the length is just right, everything is perfect. But when you come to put it in the piece alongside the other elements, it usually turns out that it is too long. So, in the context, it has to be shortened. And I do that by just listening over and over again.

NH [Laughter]

I *Nina, do you find that when you're splicing things together from individual sequences and laying everything together on the dubbing stage, things shift?*

NH Yes, massively. Obviously, the work that I'm doing is often very detailed, the minutiae of one particular element that's going to be in our soundtrack – voice, breath, and lipsmacks etc.. But it is also the case that I'm working on a very short segment in time, for example, across one sentence. And it is so valuable to wheel back and (A) give yourself the longer run-in to hear it in context, and (B) be aware of how it's then going to sit with everything else on the stage.

You need to think space wise in the frequencies, but also the arc within the entire scene, you don't want to over egg the pudding. You always need to leave somewhere to go, and you want to find the ideal high point within a scene, because each will have its own arc.

TW Yes, that's exactly the same as in my work.

NH You want to get that pay off where you need it really.

TW Yes. Totally.

Letting Go of the Smaller Details for the Wider Context

NH I used to be very, very focused solely on dialogue, whereas I'm now very much thinking about the bigger picture. It's not even just the sound, it's the sound and the picture as well – it all needs to work together in the way that it's finally going to be heard.

We have to take it in turns – effects, dialogue, and music. It's no good if we all do a gasp, then there's a musical sting, and then there's some big effects. Somebody has to take this moment [laughter]. Or we have to stagger, it just doesn't work to have everything all at once.

I *I'm interested in this question of quality and of clarity, we have the tools and technologies to record crisp, clean and clear, but actually there can be something valuable in that richness of texture in dirtier sounds?*

Clarity in Sounds/Texture

TW In the original history of Electroacoustic Music people are taught to make these very pristine recordings, very clean. But I've never really had that approach. You record something in the street, and it has all of this dirty stuff in it and that's interesting, you can work with it. I'm not necessarily interested in trying to isolate something from its context. It may be that there is something very interesting about the 'other' things in the sound. Sometimes you want that extra grittiness of the material because it's part of what your piece is about.

NH   Yeah, certainly for me it's that grittiness that is usually the bit that sells it. I'm speaking mainly of ADR rather than things that are recorded on set, where you might hear camera noise or a generator or sounds that really give away the fact this is on set [laughter]. The temptation when you're recording in an ADR theatre is to think "the original sound from set was not useable due to disruption, so we need to get this clean". But it's actually more of the case it just needs to be believable to glue back into their performance. So I'll always sacrifice getting a 'pure' recording of just voice, if their performance is better with movement. If that's throwing themselves off a mountain, they throw themselves across a room, or if they're hanging upside down, they hang upside down. You can even hear in someone's voice whether they're smiling or not.

It's not just the voice, it's in the whole environment somehow. If it is tense, people are not breathing as much, or they're not moving as much, and all those tiny little cues give you the detail of actually what is happening in the room at the time [laughter].

TW   I can relate to that, as one of my early electroacoustic pieces, *Red Bird: A Political Prisoner's Dream*[23] was based on working with voices and animal sounds and trying to make what appeared to be machinery out of very simple looping techniques. I found that I could only really get away with this by throwing in some bits of actual machine sounds recorded in factories. It was the background noise of these workplaces which helped lend a reality to my mechanically edited voice sounds.

Context is very important, someone thinks they're listening to a bird, but through a very subtle shift in the background you can radically change how that sound is heard.

NH   That's really it. One of the things that I really loved in your work was how sounds would transition. At the beginning I'd be very familiar with what the sound may be and then it transitions into a different type of sound, and it's interesting that you can do that not just by manipulating the sound that you're hearing, but the things surrounding it.

TW   Absolutely.

I   *I wonder whether that brings us back to talk about Nina's 'Mowgli' clip with these call and response voices, and that shift of environment and this interesting balance between the anthropomorphised animals.*

Voice and Performance

TW   There were a set of talking animals, and there were a set of non-talking animals (like a chorus), then this invisible character of the orchestra enters. As the scene proceeds, the animal voices appear to become part of the orchestrations, another layer that interested me a lot.

NH   It was so difficult to make the talking animals believable, you have to believe that whatever is coming out of their mouth is genuine at any moment, and to actually transition from quite lengthy dialogue scenes into a very animalistic place was quite tricky.

You need that connection to their performance, we certainly got somewhere in the end where we handed over at the right spots. It was about trying to get them in the same family so that they were believable coming out with the same voiceover. It's about fudging the transition, and by the time they've gone on with the story, you've sort of forgotten what you heard previously in a way, and it makes sense as an arc, but maybe not standalone. If you played that chunk next to the other chunk, it's the bits in between that's allowed it to work.

TW You can use people's focus to actually get away with things which you yourself are not perhaps so satisfied with.

NH You never see it [laughter].

Decision Making

I *I wonder whether we could talk a bit about scale in the sense of how you record something and then how you represent it on an array of speakers, the sense of scale and proximity? When you're recording voices and transforming them, you can en-large them and contract them, but are there limits because you want the voice to remain at human scale?*

TW There's no external construct, there's no picture constraint, there's no narrative con-straint, it's up to me what I do with that material and it's what I want to express. I don't do it arbitrarily, I think about whether I want a sense of dislocation or do I want a sense of realism? Do I want a sense that we're listening from very afar to many people talking? Or are we in the middle of some strange scene where all of the voices are whirling around us? It's my choice to set the sonic stage, so obviously I'm concerned with scale, but it's entirely up to me what I do with it. It's not something that's a constraint in any way.

Scale/Panning – Movement of the Voice

NH Alfonso Cuaron[24] in *Gravity* was very keen that you hear everything around you so George Clooney's voice would go through all the speakers and come back again, he felt that that was entirely right. And it's weird how far you can push those elements, because often if you hear a single voice coming from behind you it makes you turn around and think somebody in the cinema is speaking. So generally, our dialogue is in the centre speaker, and there might be a tiny bit of panning.

The difference is the scale, and it's when you're creating crowds or audiences, or battle sequences where you have multiple people, and you want to envelop the audience and make them feel like they're in the centre of all of these people. But it's all very much to do with the story again. Do we want to hear just one voice in the centre? Do we want to hear more spread to it? Do we want to hear them close, distant? The conventions as to whether their face/s might be filling a massive cinema screen, but we don't hear anything coming out of them, this is the same as we did in *Bohemian Rhapsody*[25] when they first came on stage. There were close shots of each band member, nervously breathing very heavily before the set started, and we chose not to put the breaths in because it kind of took away from the scale of Wembley Stadium. What we wanted to hear was that we were almost inside their heads, but not hearing themselves, and taking in the enormity of the scale of the audience. The most important sound at that point was the people cheering, waiting for them to start, the bit of silence before they started playing the music – that was the important sound.

TW   There's obviously a whole psychology of sounds – in front and behind – although technically speakers at the back and speakers at the front are exactly the same, psychologically they're not. That's another interesting thing we can play with.

NH   We often find we'll put all sorts of sounds into the surrounds and atmos over our heads and if it's atmospheric sounds then it's fine, you don't mind hearing a car going by, or wind in the trees or even crowd voices, because you wouldn't expect to look in that direction. But as soon as you hear something that you feel you should pay attention to, something singular, or that is not normal you will turn to it. It is unique pretty much to the voice and individual voices that it does not work very well [laughter].

TW   Yes, yes.

Ear Tiredness and Listening Fatigue

I   *Could we ask you about collaborative vs lone working?*

NH   So much of my work is alone, and when I'm not recording I'm listening. When working on software alone I feel like I need more frequent breaks to reset my head, and get fresh ears. It's so easy to go down rabbit holes with sound and convince yourself that something sounds fine, but it only sounds fine in the context of the last thing you did to it rather than the context of where you started. And that is something I'm constantly having to check myself over. If I finish work on anything late at night, particularly something tricky, I will never send it to anyone until I've listened to it in the morning with fresh ears.

TW   I can definitely relate to that. My experience is that I can't work on sound in where I'm essentially listening and judging for more than about three or four hours in a day. After that, I'm wasting my time because I can't, I'm not making judgments properly, because I'm not really listening. It's very interesting that you have the same experience.

NH   I do find it's different when you're collaborating with a team, when we're on the mix stage there's lots of people who are having different creative inputs and we're almost handing the baton around. We're all switched on the whole time, but you're not always doing exactly the same thing.

TW   Yeah, yeah [laughter]. I just want to say that I think it must be on the one hand, very gratifying to work on films which have enormous audiences, and everybody hears your work, but at the same time, incredibly frustrating because if the sound is designed well, they just take it for granted. They'll talk about the pictures and the story, but nobody notices the sound design, so it's a double-edged sword really, you have this fantastic audience, and nobody notices what you did [laughter]. While, I have a tiny audience [laughter] but they (hopefully) notice what I'm doing. So it's very interesting the different kind of role you have in the world.

NH   Very much so, I think it's one of those things we always say from the dialogue, ADR and Post-Production recording perspective. If I've done my job well, I'm invisible and everyone believes that is exactly as it sounded when it was recorded on set and performed live.

TW
&   [laughter].
NH

5.3 George Vlad with KMRU

KMRU and George Vlad both work extensively with field recordings, with practices built around a shared sensitivity to place, location and the identities of sounds within an environment. Their shared backgrounds emerging from electronic music into more field recording and soundscape composition, was revealed through our interviews. It is also the common respect for the sites and places in which they record, and an engagement with the people who live and exist within these spaces, which marks out a synergy between their practices.

Listening Excerpts Shared

To stimulate the conversations, we invited each participant to share up to ten minutes worth of examples which exemplify their practice.

George Vlad:
   Intense dusk chorus in Bornean primary rainforest[26] – 10 minutes, 56 seconds

KMRU:
   Continual – 6 minutes, 5 seconds

Speaker Key:

I   Interviewer
GV   George Vlad
KM   KMRU

Responses to Shared Listening Examples

I   I wondered whether we could start with your reactions to the works that were shared with you?
GV   The first thing that struck me with KMRU's composition[27] is that it made me feel things. There's a lot happening with multiple layers, similar to my experience of listening to natural soundscapes. This composition was a journey and intense from the very start. There were so many elements that were carefully mixed in and out to create a rich soundscape. So I could relate to it instantly.
I   KMRU?
KM   One thing that has struck me about George's work is that I can't work out whether the piece is composed, or if it's an actual field recording!? Because there is this repeated motif of the bird which sounds like a synthetic sound! Yet the mix and the layering feels very organic. Is it a composition made with field recordings? I'm still trying to figure it out!? [laughter].
GV   This is beautiful and more than I could have hoped for. The track I sent is a recording I made in Borneo, I didn't add anything or process it in any way. It is still probably one of my favourite sounding places in the world. The sound that you mention is not actually a bird, it's an insect – a cicada. There's so much happening in all kinds of dimensions, it evolves over time and there's so many frequencies and layers to it. Everything's happening in this one space, but at different distances from the listener.

Sometimes my recordings can be listened to as compositions. What is most important to me is that when people listen they feel something, they have questions, they have a reaction.

KM It's a really beautiful recording.

Context and Site Specificity in Recording

I *Do you think it's important that there's a connection between the sounds that you record, and the original location where they're recorded?*

KM The idea of location has become more and more important to me over time. Since I moved to Berlin, I haven't been doing as much field recording because the territory is new, and I'm trying to attune myself to the new space before I get my recorder out.

GV I can totally relate to this experience of being in a new space, you have to actively listen and analyse all the time. It takes me a few days before I am comfortable to record the scene in a new location and be confident that I'm recording what I want, instead of worrying that I'm missing something that may or may not be happening over there!

Location is obviously significant for me as I'm often travelling long journeys to get to specific locations. But there is a danger if you focus too objectively on a given location, country or space. When I got to Borneo, I felt it would take me decades to record everything and to clearly portray this location. And the danger with this mindset is that you fail to focus on what is important, which is the experience of being there, and being immersed in the soundscape.

Sometimes I share recordings without giving much information and it's funny to see people's questions – "was this recorded in X country?" Or "is this sound Y animal?" And I always say, "don't worry about that, just listen to it". If you become too analytical, you lose the focus, and the whole point of listening.

Allowing Sounds to Be

I *People always want to identify the source rather than listen to the sounds made! Do you think that's where the compositional aspect comes in, and you might start to intervene and layer up soundscapes in order to retain that focus on listening?*

GV As soon as you start to work with recordings from different places, it's more difficult for someone to say, "Oh, that's the sound of W country, X place, Y space, or Z animal". If you do a lot of processing, the listener may say, "Okay, this is just music. It's something that I like listening to and I'm not going to think about where, or when, or how it was recorded. I'm just going to listen to it", and that's the purpose of music and of sound art in the end isn't it?

KM Lately, I've not been carrying my sound recorder as much as I used to, instead I've just been in the field listening, and sometimes sketching graphic representations of what I hear. I've been thinking of the outside as a composition in and of itself. For example, on my balcony, there's been this constant droning sound and I'm not focusing so much on trying to think about what is causing the sound, but how it creates an atmosphere and appreciating it for what it is.

GV That resonates with me, my interests also include ecology and ornithology, and there's a tendency to be analytical, part of my brain is always trying to identify sound sources. It took me a long time before I could disengage it so as to appreciate the beauty or the ugliness of the sound textures happening.

Good and Bad Sounds

I   *Do notions of "good and bad" sounds direct your practice?*

KM   I try not to be judgmental. From what point of view is something considered good? As soon as you ask, "what is a good sound?" you immediately adopt an aesthetic approach to soundscape, it is very problematic. When I'm outside, I'm totally open to the experience of sounds and I don't make a judgement on what I'm intending to record. I just let things come to me. Lately, I've been recording electromagnetic sounds, which are very noisy and interesting, but somebody else might consider these as very harsh and 'bad'.

GV   Working in game audio and film audio, you think about clean and dirty sound. If you can record an animal up close and focused, with nothing else around it and no other interference, that's perfect! It's the Holy Grail for use in sound design.

I started off by recording ambiences and then I realised that they were not really that usable as there's a lot going on in one scene. So I bought a shotgun microphone to be able to differentiate between clean and dirty sounds when recording. But that's when I realised that I was ignoring certain things because the sounds were not clean or focussed enough. It took me many years before I circled back to recording ambience and not worrying too much about this notion of clear isolated sounds.

A formative moment for me came when I was in the Amazon rainforest, I could hear a stream babbling nearby but there was no water, just humidity, and so I went to investigate. It was actually a big log that had fallen to the ground and was being eaten from the inside by grubs. But they made this beautiful, disgusting sound like clicking and water dropping,[28] and I was so taken by that sound, I couldn't move from there. So, I did some recording and there was wind noise, birds singing and insects all around. This was not a clean sound, it's quite the opposite, but it's still one of my favourite recordings because of the mixture of all these textures. I can listen to it and disregard the fact that it was a bit windy and the microphones were a bit noisy, I can just listen to it for the aesthetic value of that sound. And that was a big achievement for me because it helped me to transcend notions of clean/dirty, good/bad.

I   *There is this tendency to think about nature sounds as good, and urban sounds as bad. Do you feel that that is something that reflects in your practices?*

GV   When I started out, I was trying to get away from manmade noise as much as possible. I was so focused on recording the sound of pure nature on a trip to Senegal,[29] that I missed out on recording a lot of other beautiful things, such as the village dances, and interesting celebrations that took place. I bitterly regret that now.

I now really enjoy capturing cityscapes. I like to listen to the cycle, which is very much like a dawn and dusk chorus. In the morning people get up and go to work and you have this rush of activity before a lull in the daytime. People then come back from work in the evening, and you get all this racket and noise, before it gently fades away as people go to sleep. It's a fascinating thing to listen to. I really like the sound of distant human activity, the signs of human life.

KM   I also began with a focus in nature, capturing different bird songs, insects, etc. But I became more and more interested in the natural soundscape of the city, because in Nairobi especially, there's so much happening and so many textures emerge from the city. For me, I noticed a huge difference when I moved to Berlin. Suddenly I found myself in a totally different sonic environment, much quieter, and I have had to really explore to understand how this new space sounds.

GV   I can definitely see the impact moving from Nairobi to Berlin. I was recently in Nairobi, and left my recording equipment out for 14–16 hours. There's a lot of dogs barking at night, and a lot of chaos and traffic in the early morning. In some ways it made me think of growing up in Romania, where there are a lot of feral dogs, and everyone has a guard dog. As soon as the human element dies down a bit, the dogs take over with a lot of barking to communicate with each other. And that's how I felt when I was in Nairobi listening to that soundscape.

Time and Listening – Duration

I   *On this topic of listening to the city and long form cycles, I'm interested in your approaches to time and duration.*

KM   I realised that when I was making electronic dance music, I was focusing so much on the timeline and the grid of software that I'm using.

But when I started field recording, I found myself outside for longer and longer, recording greater durations and then bringing them back in the studio. This led to a clash in working methods, I didn't want the recordings to conform to this linear idea of grid time, because what is happening outside is not fragmented into such a timeline, it's fluid.

So, this tension became my compositional approach, improvising freely with instruments alongside a field recording of around 20 minutes in duration. These durational pieces would gradually evolve over time, and this process of working pushed me to be very aware of what I'm doing with my instruments and synthesisers, and to listen more.

It then became an experiment with audiences, trying to see if people would really spend 20 minutes in a club context, sitting down and listening?! Eventually, I think people got used to listening because I was foregrounding the field recordings, they were not subtle backgrounds but very clear, and I think people could relate to those sounds from their everyday life.

GV   This is very similar to my experience. I attended a sound design symposium in 2012 with Chris Watson[30] and he was talking about how contemporary culture is conditioning us to pay attention for shorter time periods. We're moving from pop music that lasts five or six minutes to tracks of two or three minutes. Everything seems to be getting shorter and shorter, and people seem to be moving away from allowing things to develop over time.

But when you think about sound as an experience, short durations don't allow enough time to understand what's happening. It will be a waste if you share just two minutes or thirty seconds, as there's not enough time to establish the scene or appreciate how it evolves.

Nowadays I record for 24, 48, 71 hours to take it all in, and then condense that down into a few hours of edited track to portray a space, or a species, or something that I find exciting.

Being in the Space and Ethical Responsibility

I   *Living there or being there for longer gives you a much greater picture. And this is another thing that both of you talked about in our individual conversations – the idea of inhabiting a space, being there yourself, and also conversing with people who live in that space and have always lived in that space.*

KM   Yes, I love it if there is a potential to go back to the places that I've recorded previously, I want to hear if something has changed or if it's evolving. I always want to stay longer!

I did a project in Kibera,[31] where I spent a one-week residency getting to know the people and how things worked inside the slum. I wanted to really understand and get to the roots of the place, and I worked with a 'sound guide' who told me the best places to record. This aspect of the space and people welcoming you is very important, as it's important not to be coming in as a as a sound invader, just stealing sound recordings and going home.

GV   I actually had an offer to tour Soweto in Johannesburg, which is a huge township and very poor and underdeveloped. And I decided against it because I would have only been going in for a few hours, maybe doing some sound recording, and then going away, and I didn't feel okay with that. I would like to spend a week, like you did in Kibera, to go there and speak to the people and try to understand their way of life, and immerse myself in the landscape. Even though I'm not sure how I'd approach it, obviously, I feel like an intruder sometimes in these situations, but having a local guide, who's grown up there or knows a lot about the space is definitely very helpful. It's a way of getting permission to access a place and record it or document it, or even to experience it. I find it also opens my mind, eyes and ears to things that I would ignore otherwise.

To Be Part of the Recording? Or Not

I   *I wonder whether you could talk about how you feel about the presence of 'you' as the recording artist in the space, when you go to a space, how do you negotiate this idea of you being in the space, your presence as a recordist?*

GV   It was very interesting to be on a sound recording / sound art workshop in South Africa,[32] and to observe other people's practices. One of the attendees was recording herself walking in a space and then pointing the recorder to various things, and then talking into it. At the time, that was completely counterintuitive to me. I was thinking, "How are you going to work with that? What are you going to do with these recordings?".

In my work for filming and for video games, it's very important not be present in the recording itself, it's important for nature and wildlife to get back to a normal state and to be going about their business as usual. Being there would inevitably affect the soundscape as creatures will be alarmed by your presence, smell, and by however you're affecting the environment. They will still be making alarm calls for maybe 30 minutes, or an hour after you've left. This is why, when I go somewhere few people generally go, I set up an unattended recording rig that I can leave there for two, three or four days. But that's obviously still not 100% perfect. My equipment being there will still affect the soundscape, there's going to be a bit more interest from certain animals, others will be shyer, and they'll try to get away from it.

On my trip to the Amazon in 2019, I took a canoe ride at dawn on a tributary in the rainforest and it was impossible to not be in the recording. There's the movement of the canoe through the water, the sound of the wooden oar hitting the wood of the canoe, the sound of the oar hitting the water and splashing, my own movements and the boat creaking. Listening back, it's probably the most alive of all of the recordings I made on that trip, because otherwise they're very clinical, as they're recordings of a space, and there's no human element. And so, there's something that I like about me being present to a small extent in the recording itself.

KM   When I started out, I had a naive approach. I was just excited about recording and didn't know much about the practice. I was free and open. I started off with a hand-held recorder and I was usually always there holding it (occasionally I might leave it somewhere, if I knew it would be safe). But I like the fact that I also am part of the recordings – the sounds that my body makes, hand movements that get picked up by the recorder, sometimes I don't really alter them, because I'm also thinking of how the recordings will be used in my compositions.

If I want to be very clinical or use the recordings as clean extracts, I know I can process and remove these noises, but I often let everything be present in order to be honest to the fact that I was there while recording.

I understand that nature will always react to the human presence, but because I'm doing a lot of urban recordings, I have to think about how other people will react. In Nairobi, few people knew what I was doing or questioned when I used the handheld recorder. As soon as you change up microphones, use windshields and have longer cables, it starts to draw attention and people will act differently or change their attitudes.

Memory, Experience and Recording

GV   In 2018 I went to Gabon, and we hiked for days to get to this special clearing in the rainforest that attracts elephants and an incredible amount of wildlife. I had to make a choice, I could either camp there and experience this soundscape for myself in person. Or I could set up a recording rig to preserve the undisturbed condition of nature and come back to collect the recording rig the next day. I couldn't do both.

Eventually, I chose to leave my rig out and I got an incredible recording. But I still question my decision because I would have loved to have been there myself. I might try to go back one day just to be able to experience this in person.

How do you feel about this conflict between, desire to be in a space to experience it for yourself, but also to record it?

KM   I believe that I'm an active agent in the recording process, a sound person, and as such I'm a part of that recording and how it turns out.

I find my memory of the experience very important. If I listen to an unlabelled recording I made (e.g. Recording 001), I can instantly remember where I was and what was happening, I'll maybe even picture something that happened. If I'm not present at the time of recording, then I might miss out on understanding the significance of elements in the sound file.

Sonic Focus vs Openness

I   *To what extent are you controlling and limiting what it is that you're recording?*

GV   There's a temptation to try and record everything (or to record in ambisonics[33] so that you cover everything). But you have to think carefully about the listener because it can be overwhelming. By making decisions you can often make the listener feel something more. I'd often rather listen to a stereo soundscape that was carefully framed in terms of stereo technique, positioning and location, than a full 360° ambisonic soundscape.

KM   As with George, it's mostly dependent on the setup that I'm thinking of for the final presentation, or perhaps what tools I have access to and familiarity with. I have a pair of binaural microphones that I wear as earphones to record an impression of my own sense of space, and what is happening around me while I'm walking through the streets. I'll also carry a handheld recorder to capture the space itself. But I tend not to think so much of what's happening around me spatially in the recording itself, at the end I'll usually listen back to it in stereo.

I   *So it's more about providing focus on a particular part?*

KM   Yes, it's like having this panoramic perspective on how the space sounds overall, and then zooming into focal points.

GV   I love listening to stereo recordings. Sometimes they're even more immersive than 360° or Ambisonics, because with Omni[34] mics you can capture a wide stereo field and you get a good understanding of what's happening and a sense of space. But most importantly, that framing is directed by the recordist's perspective – in listening you're learning about the space through their lens/ears. The agency of choosing the location and how you record is very important. Otherwise, it could be done by robots and you end up just with this grid that records everything. Hopefully, that's not the future.

Recordings as Performance

I   *Do you imagine recording as a completely transparent objective thing that will translate the sounds that come into the microphone, or do you see it as a creative act? That process of intervening.*

KM   It's definitely a creative act for me. I don't think you can ever disconnect yourself from the subject, and this notion of a 100% transparent recording is just not possible.

GV   Yeah, it's also the same for me, it's very pragmatic, the creative process of being there and hitting record.

I   *This idea of being there and able to respond to what's happening in real time is quite interesting. You're able to capture more vitality. Is this almost treating recording like a performance in a way?*

GV   I did feel like the canoe I mentioned earlier was a bit of an instrument that we were playing as well as gently guiding towards a certain meander in the tributary. I was paddling and gently moving whenever I simply couldn't stop myself from moving anymore. I worried that I would ruin the recording, but it just ended up sounding beautiful. It sounded like people on a canoe going up and down the tributary being quiet, not being impossible to hear. There's something in that human warmth and presence that makes this recording a bit more enjoyable and fun to listen to, and interesting.

KM   This reminds me of a recording I did in Paris at the GRM studios. The studio is four floors below the ground, and they have these great resonant frequencies in the stairwells. I set up a microphone and performed in the space, getting as much sound as possible from moving up and down, hitting the rails and throwing things down the stairs, in order to hear how the space resonated in itself. This exploration is one of the most interesting things that I've done recently.

GV   Yes, that seems a much more organic approach than using impulse responses of a space. You record something that's alive, something that's human, not just clinical bursts of noise. I like that much more.

Humans within the Soundscape

KM   George, do you record the voices of people speaking while you're out field recording?

GV   I'm fascinated with languages and whenever I go somewhere where it's not just English, French or Spanish, I like to record short bits of the local language and people singing, talking and going about their daily business. It's not just treating the voice as a source for documenting the place or a particular acoustic space. It's something alive that I can relate to, even if I don't understand the language itself.

  I used to not care much about the human element, and I very slowly moved towards recording all kinds of unusual music and people. We are surrounded by humans (hopefully), and it's so easy to relate to, it's part of our upbringing, part of our nature.

KM   I feel there's so much presence and emotion which comes from the human aspects in our recordings, and it brings things that we can all relate to. Letting the human be part of the recording and not trying to omit it or change or remove it, is something that I appreciate in my recordings.

Chance Processes

I   *How important is chance and the role of the unexpected in the work that you do?*

GV   Setting up a rig and leaving it somewhere for a few days is all about chance. Listening back to what I recorded is then like opening presents on Christmas day! Many of my best recordings are made that way. I captured a recording of rain in the Namib desert by putting microphones on the branches of a tree, and when the soft rain eventually came, the drops hitting the wood made it sound almost like an instrument being played. The wood was so resonant it sounded like a marimba.[35] I could never plan for that, it never rains in the Namib. So it's just a question of doing it enough times and eventually chance brings something into my recordings. I love this element and I'm trying to disconnect from trying to control everything.

KM   For me, it's more about learning from how sounds operate in the world outside and finding ways in which I can replicate or integrate those features into my music. For example, I learned to embrace the unexpected and irregular from listening to the environment where things are just happening randomly in an unplanned way. So I don't use the metronome or strict measures in the timeline, I just have a very gridless space to work in.

  I'm also drawn to tools or functions which are very free and spontaneous, so that I don't know what will happen when I trigger something. I find this chance encounter exciting.

GV   I definitely got that feeling from your composition, there was so much micro randomness happening. Obviously there was structure overall, but there were elements that didn't feel they were attached to a grid or anything.

Do Mistakes Happen?

I   *Is there such a thing as a mistake?*

GV   Maybe if you end up losing all of your equipment to elephants, that's probably a mistake [laughter]. Or maybe even then it isn't. I was experimenting with hydrophones[36] a lot a few years ago, and some of the best recordings I made were from the interactions of me moving and touching the microphone cables gently, and getting these weird textures that I've no idea how I'd be able to get otherwise. I got them by accident when I was trying to be quiet! [laughter].

KM Mistakes are just unlucky experiments. It could be that they don't work out or something breaks, but other times, they can really end up working well! I've also been experimenting with hydrophones, mostly freezing them in ice and listening to the cracks. I never know if my microphone will break, but then perhaps that will make an interesting sound [laughter].

I know that many sound artists find aircraft very annoying in field recordings, and it's a shame that so many things have already been normalised and considered as mistakes. Obviously, it's no good if you're trying to capture clean dialogue for a film and an aircraft goes over. But if you're out recording the sounds of the city, why would you choose to ignore one of the most present and iconic city sounds?

5.4 Steve Fanagan with Annie Mahtani

Steve and Annie have a shared interest in constructing impressions of space and spatiality. Working with multiple microphone recording approaches to capture sound and layering, they capture different perspectives on space working towards developing deep and multi-layered sonic worlds that draw on impressions of real places to convey affective experiences.

Listening Excerpts Shared

To stimulate the conversations, we invited each participant to share up to ten minutes worth of examples which exemplify their practice.

Annie Mahtani:
   Inversions[37] (extract) – 2 minutes, 11 seconds
   Racines Tordues[38] (extract) – 4 minutes, 40 seconds
   'Round Midnight[39] (extract) – 3 minutes, 15 seconds

Steve Fanagan:
   The Little Stranger[40] (extract) – Two versions of the same sequence, the first is the pre-mix, with temp VFX and score, the second is the final mix with finished picture and no score. From the opening 20 minutes of the film – 2 minutes, 24 seconds
   The Little Stranger (extract) – Final version of a flashback sequence – 4 minutes, 3 seconds

Speaker Key:

SF Steve Fanagan
AM Annie Mahtani
I   Interviewer

Immersion in the Space of the Scene (in *The Little Stranger*)

AM It was such a marked change in experience having the two versions of the scene from *The Little Stranger*.
       With the orchestra it felt saccharin, and as if it was pushing me towards feeling a certain way. However, when I listened to the version without the orchestra, I just had a sense of being in that world.
       It was so much more powerful relying on the silence and the natural reverberation in those rooms. The significance of each moment was magnified. As a listener I became much more attentive to the tiniest sounds, like footsteps, breath and the clock chiming.
       There was tension and I knew that something significant was going to happen, I found it much more ominous.
SF   The final decision about which version of the scene to go for, was made during the final mix by our Director,[41] and by his Editor, Nathan Nugent.[42] By allowing the character Faraday to be in the house, and for it to be the sounds of the house that surround him, you let him luxuriate in that moment, and create that personal connection between the character and the space.

In the sound design we were trying to capture the experience you might have walking into a space that reminds you of what it felt to be there 20 years ago. It hopefully puts you into the head, makes it subjective and leads you into Domhnall [Gleeson]'s[43] performance and the work by the production design team, allowing those elements to speak and highlight their significance.

In the sound department, we're always trying to build scenes with the idea that the music may not always stay. Often in offline [editing], music is used to support something that sound hasn't yet had an opportunity to help with. And if sound does the right job we might make enough space, or create a mood, that doesn't require something melodic.

I fully give all credit on the decision to Lenny [Abrahamson], and it's an example of his confident filmmaking, which I really admire. It's a great thing when a director is confident in their craft. Because that confidence gets passed on to the rest of us. For example: Stephen Rennicks[44] who is Lenny's long-term Composer, is often the first person at a mix to say, "maybe we don't need music here". And that can be quite rare. It takes a certain confidence to not feel like someone is rejecting your work by choosing to take it out. And being part of a team like that is really exciting for me.

Naturalism and Loudness – Film Sound vs Life Sound

AM   If I think of the moment when Faraday touches the mirror, you can feel the texture through sound. When you have the orchestra, you can still hear it, but you don't have that same direct sensory relationship with it.

Did you change any of the sound when they took the orchestra out?

SF   We did reshape slightly, because suddenly we had more space. For example, we were no longer trying to push the levels of the rain so it could be heard above the music. We could just place the rain at a level that feels naturalistic, or slightly hyper-real. And the same for the Foley on the mirror.

When you have to push sounds to an unreal level – so they can be heard through music – it can become less impactful for an audience and maybe less naturalistic. It feels like film sound, rather than life sound. So, the rebalance was probably more about not having to push things quite as hard.

AM   It's interesting this balance between naturalistic levels and the hyper-real, which perhaps reflect the mental states of the characters.

SF   Absolutely. I think what you're trying to establish at this point in the film is that this house has some sort of magnetism and hold over the character, and so his experience is almost as if he's rediscovering an old friend.

He doesn't touch this mirror in the same way he'd touch another mirror. And by heightening the sound we hopefully communicate to the audience subconsciously, that this is an important object, and what the character is feeling.

Sound Materials & Building Structure

AM   Do you mostly work with recorded or synthesised sounds?

SF   It's really a mixture of everything. In *The Little Stranger*, we had access to the house that they filmed in. It was a really rundown stately home in Watford, built in the 1700s and added to over time. Nobody had lived there since the 1940s so, it was in quite a state of disrepair, which was perfect for us because everything creaked and squeaked, and moaned and groaned.

We had a few days to record during the shoot and used lots of different microphone types – contact mics[45] and stethoscope[46] mics – to see what sort of resonances we could get from the building. Unfortunately, the actual house wasn't amazing for atmos[47] recording because it was so close to London. You could hear the hum of the M25 and lots of helicopters.

We also recorded in a couple of other places, such as a stately home in Enniskillen.[48] There was this amazing attic in that house that had very creaky floors and lots of loose floorboards. And when you slowed and pitched it down, a lot of that became quite atmospheric.

As with any film, you're trying to record as much original material as you can. And hopefully some of that source material inspires you. I discussed with Lenny, that the house has to bend towards Faraday over time, it has to become connected to him almost like it's haunting him. And those creaks became brilliant for that.

We were processing sounds, but I loved that the sounds were anchored to the house. Being from that location they have an immediate attachment to it. You're starting with material that has the correct natural acoustic, and so both the sound effects and on set dialogue are both living and breathing with the same acoustic space.

AM   I have lots of notes about floorboards and creaking in the reflections I made while listening, they're integral to the scene and really emotive. It's what provides that tension.

And that really significant moment, when he snaps the acorn off of the mirror, and you've used a massive augmentation of the gesture. I was struck by how, it's a tiny event but [laughter] clearly it has the biggest impact on this child's life.

SF   That is *the* key moment! But if you take what we did in isolation; it seems totally ridiculous. If you jump in directly to that moment you might think it's totally over the top. But, we create little mirrors or little breadcrumb trails leading into that moment. For example, perhaps he steps on a floorboard or opens the door and we use a quiet version of that large gesture as part of these smaller sounds.

The idea is to communicate that this sound gesture is so ingrained, and it's followed him through his life. It's a moment of schism, the shame and humiliation of getting caught and how his mother reacted to it, he's never really broken away from that. So, the moment has to be larger than life, and there's great licence in that.

But it only works if you have the right story, visual material, editing, direction, to allow you to develop these things over time.

AM   That gentle brushing of finger over the mirror, which is so delicate stands in complete contrast with that loud moment and I thought it really worked in expressing the enormity of this relatively tiny gesture. I felt for this poor child.

SF   Well, that's lovely to hear because when I'm working I'm just trying to use my sensibility to respond to what's in front of me, trying to figure out how sound can express feeling? For me there's nothing else that's important. I don't really mind what sounds I use, so long as they evoke.

Sound and Feeling – (in *Racines Tordues*)

SF   That's what I find really interesting in your work, Annie. However abstract your materials might be, the sounds you're using 'feel' real. They feel like they're connecting to something inner, to the rhythm of my pulse, or the sensory experience you have in an emotional moment.

Perhaps there is not a massive difference between what we're doing? We're both thinking "how can I present this sound, so that it evokes a feeling that I want to elicit in the audience?"

AM Wow, thank you. We often rely on programme notes to be able to give guidance to the listener, but it's lovely to hear that that's coming through without any knowledge of the pieces [laughter]. The work I shared [*Racines Tordues*] was written for my sister after she went through a really difficult period. And so, it is about going into a very, very dark space, but using that trajectory to transform and to untangle so that it resolves at the end.

SF I was totally fascinated. I didn't realise that there were field recordings involved, I felt I could hear textures that had organic qualities, but when I first tried to formulate my response to it, I kept coming back to this idea of subjectivity and feeling like I was inside something.

It was highly sensory for me and very emotional. It was dark and overpowering, very visceral. Two of the pieces started calm, but evolved into something much more frenetic, claustrophobic and disturbing.

And I kept coming back to the idea that it was an awakening of some description. A potential realisation of apocalypse perhaps? I wake up and the sun is coming up, then as the moment transpires, the fact that the sun comes up has basically caused a fire that is engulfing my experience.

Composing Spatiality & Point of Audition

SF When I went back and listened repeatedly, I was able to reflect upon the technical elements, in particular the spatialisation. I was struck by how the perspective as a listener is totally first person. I felt this piece was me. I wasn't listening to it; I was inside of it.

In film work we're always trying to identify and articulate this 'point of listening', to figure out where we place the audience so that the experience tells a story and does something emotional.

AM It's really nice to hear that you felt completely immersed and had a strong sense of space. Space is one of my fundamental preoccupations, whether that's in a concert hall or listening on headphones.

My approach to space comes from my experience of Loudspeaker Orchestra per-formance, manipulating sound around the audience live.[49] That final performance really informs how I approach space in composition, because I'm always imagining that bigger concert space and how I can articulate sounds spatially in the concert hall.

AM 'Round Midnight is made using recordings from the Amazon Rainforest, which have not actually been manipulated a great deal, but they're just so complex in and of themselves. They've been layered, but there's a complexity in them which just gives them a hyper-reality. In some cases, they become quite abstract. There's so much going on around you when you're just sitting in the jungle and when you don't have the context of it, it could be anything.

Space is integral to the piece because it's recordings of huge spaces that have got so much content in them, and spatial content, as well with things moving around you, closer and further.

Balancing the Truth

SF   If you're composing in a 16-speaker studio, but the final piece is going to be in a gallery space or an auditorium space, do you get the opportunity to mix in that space? Or at least listen and then tweak your mix?

AM   I don't often get to mix in the final performance spaces, but one of the traditions and practices of Electroacoustic Music is the performance of sound compositions over a loudspeaker orchestra. This practice, called 'diffusion,'[50] evolved as a way of translating studio composed works into a performative concert setting. It's not a case of remixing the material (the original wav file of the piece remains the same) rather, you choose how the track is distributed throughout the space and how that evolves over time. For instance, if a part of the composition feels 'distant', I can push the sound to the distant speakers at the back of the auditorium to emphasise that effect.

   I can augment the spatial features that are present within the original track, as we have various levels of height and depth of perspective to play with. I will only edit the mix if the acoustics of the space are completely unexpected.

SF   When you're spatialising live, or when you're thinking about space, is that an emotional decision you're making?

AM   In some ways, yes, because you're imagining "what space am I putting my audience in? What do I want them to feel?" And that will inform what I do with the space.

   It is about responding to the material itself. Through having the experience of presenting works in different concert hall settings, I have an understanding of how certain sounds will react in different acoustics.

SF   I can relate to this. When you go into different cinema spaces it can be quite a different experience, depending on how well maintained and/or well built the cinema is. So, you're often mixing for an ideal or optimum space, but taking into consideration that not everywhere is going to reproduce the sound in the same way.

   It's a real challenge when we think about adding reverb in the mix. You don't want to end up adding something that will sound over the top and become unintelligible when it plays in a less than optimum space (particularly dialogue). So, there is a delicate balancing of the truth. We might use a reverb that is highly engineered and loud, but refined to sound natural, and therefore has the required impact on the audience.

   Acoustics are so important. People don't think enough about the fact that microphone placement (and point of audition)[51] is not just technical, it's always emotional. And that shone through your work.

   I've listened to lots of music on headphones, but I don't recall ever feeling like I'm inside of it like this before. Your creative use of spatial audio is a credit to what you're doing, and a pleasure for us as listeners.

Sonic Focus

SF   When you go out to record is there something specific that you're looking for?

AM   I try not to have too much of a fixed idea of what I want to capture when I go on a field recording trip, because the realities of the location will always be different than you imagine and if you get too fixated on a desired outcome, I think you are only setting yourself up for failure [laughter]. So, I go with an intention to explore. Often it is this very personal process of exploring the environment that inspires the work I will eventually create.

This summer I've been working in the Brimingham Institute of Forest Research (BIFoR)[52] which is an outdoor laboratory, there's a natural soundscape but also a lot of manmade sounds, generators and pipes etc. On the most recent trip, I wanted to capture the spaces of the forest, and I made multiple ambisonic recordings at different heights, as well as using little DPA 4060 microphones[53] to record the crevices of trees and contact microphones to pick up inaudible sounds. Often, I'll do these recordings in parallel so that later I can mix between the different perspectives.

Perspective is something I find fascinating to work with.

I *Can I ask you to expand a bit more on that please? You talked about using different microphones, and that exploratory process of finding sounds but why do you choose the sounds that you do? What is it that makes you choose 'this sound' instead of 'that sound'?*

AM I think of soundscapes as a visual image, a landscape; what's happening within it, and how it is constructed. I use that scene to guide me when I'm placing my microphones, imagining that they are on a 'virtual stage'.

For instance, on a recent trip to BIFoR I placed my DPAs on a half-dead tree trunk. The height of it meant that the birds and the canopy were above me, yet I was also picking up micro-audible sounds from the forest floor. It's framed so nicely, on the centre stage.

Sounds have to be saying something to me. I listen out for a combination of natural sound qualities (already present in the materials) and future potential (what can I manipulate and draw out).

I *You're taking recognisable recorded materials and using them to represent emotional experiences. So, there is a question here about the truth and honesty of the experience, wanting to reflect that experience of being there?*

AM When I came back from the Amazon, I had 50 hours of field recordings, which I could sit and listen to quite happily. But I can't expect anybody else to listen to all of those, and get the same reaction that I have, because I carry with me the memories of what it was like to be there in the rainforest.

I have to detach myself from the recording experience, and make the decision to frame specific elements from the recording. To give listeners a time and a space to be able to listen, you have to zoom them in really quickly, and frame the overall experience so as to provide a space for them to listen within.

SF Exactly! When I go out and record, I'm often listening for little moments. Over several minutes there might be three things that jump out at me which I couldn't have planned – a snippet of voice as someone walks past, a car in the distance that does something unexpected, or water lapping on the shore in such a way that you just feel something.

I always listen. I never leave the microphones set up and walk away as it's vital for me to connect with the space – to actively listen to that moment in time.

Crafting Emotional Trajectories

SF I'm struck by the fact that our challenge might be different in terms of how we present the work, but in essence, it's the same idea. We have all of these sounds and possibilities, so how do we curate this so that it has the most useful impact for the audience?

Picking what you want to feature and where is key. One term I use is "to pull in focus", because it's very useful for me to say to a director, "I think we need to pull focus on this sound, because it will work for the story."

We record every footstep in a film, but we might, in the end, think about only using half of them or less, and it just depends on when they're dramatically important or when they might be story-wise distracting. It's all about that focus.

AM   You were saying that the ambience was difficult to get on location for *The Little Stranger*. Did you record any of the rain there? Or did you have to build a set to record on? Because there is quite a progression of rain across that sequence.

SF   Rain is a really difficult thing to record so I'll use ten or twelve individual files of different focus. If you record heavy drips down in foliage, you get a very particular sound; if you get sheet rain blowing against a window, you get a very particular sound; if you're inside a building that's well insulated, you just get a bass rumble from the rain. So, you're trying to find recordings that reflect each of these different perspectives and edit them together to fill out the cinema space, and feel naturalistic.

In *The Little Stranger*, we move into the house getting progressively further removed from the rain. But as you come towards the stairwell with its central glass dome, the rain has to become more prominent again. Having the correct rain textures was really important to that sequence. The rain in the hallway needed to feel more like a rumble because there are no obvious windows, but as you move toward the stairwell it becomes all about that additional patter on the glass dome, as well as that rumble. But it needed to feel cold listening to it.

We have such an emotional connection to different experiences of rain. If you're in bed and you don't need to get up and it's raining on the window outside that can be a very warm feeling. But if it's flipped and your alarm's about to go off and you're about to commute, it's a very cold feeling [laughter].

It is fundamentally an emotional sound to me. I think we all have connections with sound that we don't think about consciously, but if we can be aware of those possible connections that people might have, we have something powerful to articulate and use subconsciously with our audience.

AM   I can imagine rain is one of those sounds that you often give a hyper-real treatment because if you just recorded it in the house, it might not sound quite as dramatic as it needs to for the actual narrative!

SF   Definitely. It would be really boring [laughter].

Orchestrating and Arranging Sound Textures

I   *The way you describe it is almost like the orchestration of these multiple layers. Treating them as one texture that you're then shifting the balance of and modulating over time?*

SF   Yes, it's like an arrangement. And that is where my musical experience has become very useful. Our ears are always positioning us in relation to what is happening around us sound-wise. In music you think about what's available to you in space and frequency, and how that might play for an audience in terms of how they interpret a sound.

If we hear a little slap [echo] we know that the voice we heard is far away. Our ears and our brain are constantly reading acoustic information that says, "you are here. And whatever is around you is either close, in a scary way, or it's far enough away that you're safe".

I find myself really fascinated by that in terms of film because, presumably, that's what the characters are experiencing. So, if we can express their relationship to the sound source, then hopefully we are providing something that is a good experience for the audience in relation to the character.

SF   The realisation that I keep coming back to is that you just can't force an idea onto the picture. It'll just reject it.

No matter how cool your idea is, or how interesting, or how theorised it might be, the picture will just push back and say, "no, bad idea". And I think that's where experimentation comes in. We have to always be experimenting in the sound edit and the mix.

It's easy to do a working version of a scene. It's hard to do a version that gets to its essence, because you have to go down a few of those blind alleys where a couple of days of work go down the drain, because you've tried something that the picture just won't accept.

And I'm sure that's the same for you Annie?

AM   Yes, exactly as you talk of the picture rejecting the ideas, so will the characteristics of my sound materials limit and direct possibilities. You've got to understand the character of the sounds you have and know how best they will behave, either in articulation or through processing – how far you can abstract it. If you respond to the affordances of the sound, then the result tends to feel natural. But if you try and push things into places where they just don't want to go, for example: rapidly panning a low energy sound around the room, when all it really wants to do is just wallow, that's never going to be successful, and it's going to sound false.

5.5 Onnalee Blank with Nikos Stavropoulos

Nikos and Onnalee share an interest in in the performative nature of gesture, and the articulation of spaces for creative effect. Their works are richly layered with highly detailed textural components that construct immersive sound worlds transitioning between impressions of reality and greater abstraction. Listening to their sounds, one has the impression of a very direct and human gestural physicality, with an apparently direct and tangible effect.

Listening Excerpts Shared

To stimulate the conversations, we invited each participant to share up to ten minutes worth of examples which exemplify their practice.

Onnalee Blank:
  The Underground Railroad (Episode 1) (7.1 format) – 3 minutes, 1 second

Nikos Stavropoulos:
  Claustro[54] (5.1 format) – 7 minutes, 51 seconds

Speaker Key:

I	Interviewer
OB	Onnalee Blank
NS	Nikos Stavropoulos

Sonic Exchanges

AKH What were your reactions, responses, feelings and impressions after listening to the extracts that were shared?

OB Nikos, you need to compose a horror movie! Your stuff made my mind race. The music, stingers and the risers that change in pitch and shape – which evolve but are not stagnant. It's something that you might hear in *The Shining*,[55] as Jack Nicholson is trying to come through the door.

I was curious at first, wondering is there any picture for your material? But no, it's just sounds, so let me get into the zone here. I listened to it a couple of times – you have a lot of layers, there's a lot of things going on in each little section, and it's fascinating. It's very cool and different, and not something that I've heard before.

Your music is very visual and, from a film standpoint, there were a couple of sections in your piece where I thought, "Oh, this is scary!" [laughter]. But in a good way! Really cool sounds. I don't even know what you put together to make some of those things, but it was great!

NS Thank you!

OB How did you start writing your piece?

NS The starting point is always the recording. I often don't even know what to record, so I'm collecting whatever is around and aiming for a combination of either high end, gestural or more textural material. All the ideas come from the characteristics of the sound itself. And for *Claustro* – and my other recent works – I've captured multi-channel recordings as the starting point, so everything that you heard is multi-channel audio, from recording all of the way to the end.

OB   That's really cool. Not too many people do that.

NS   No, I've been exploring this for the last two or three years. There's no panning automation used in *Claustro,* all of the movements were captured in the recorded space or emerge as a result of multichannel processing.

OB   Wow, that's great. Do you have a multiple microphone set up that you move around based upon what you're feeling and vibing on?

NS   I used a custom microphone, made in collaboration with colleagues.[56] This was the second-generation model, and all of the sounds in *Claustro* were recorded with this.

It's a compact array that I can insert into compact spaces [Nikos shows an image of the microphone array on the computer screen]. The idea being to capture movements or spaces that are very small and intimate. I'm then granulating the 5.0 recording, working with GRM Tools.[57]

OB   Wow, that looks so neat, and the granular stuff is cool! I'm very impressed that you didn't do any panning, that seems very hard.

NS   Thank you.

In regard to your own work Onnalee, I listened to the whole episode that you shared,[58] with just sound, and it carried me with it. There was a strong sense of intimacy in your work, and it became quickly evident that we share a lot in terms of technique. For example, there was a sequence where it sounded like insects and high-frequency granulation and this felt very close to home in terms of the types of sounds that I'm trying to create in my work.

But there were also more gestural moments in the sequence, which made me think of other Electroacoustic Composers such as Jonty Harrison.[59]

OB   This was a period piece about slavery, and I discovered halfway through working on it that this was actually a horror movie. Slavery is a horror as these people are living in fear their entire lives. I knew that I needed to portray this in the sound, and I made the sound a character, almost like a slave catcher, in a way.

The ambiences – crickets and winds – were all recorded for the project. I put microphones up in trees and I had someone camp out and sleep in a tree for three days [laughter]. I had no limits, I panned and mixed, whatever I could do to create the effect I wanted but retain a naturalistic sense, I played things in half-time and added delays upon delays.

I used delay on everything [laughter].

NS   It was awesome! I was wondering how much you process your atmos materials and field-recordings?

OB   Every sound was morphed, every single thing. Even if they start out real, we'd record at 92kHz, 192kHz, and then slow stuff down and put different stuff in. The surrounds have a different delay than the fronts, and we would play with spreading out that delay even more and test what it sounds like.

The EQ is also moving throughout the whole piece as I don't like stagnant backgrounds in a movie. In the real world everything is constantly changing, so when we are mixing, I feel it's so important to replicate this, I say, "Hey, let's just have a dance party here and make it weird".

And that is what's great about working with the director Barry Jenkins,[60] he really allows you to be and get weird. He would often say, "You are all making it sound too natural. Let's make it sound like we are in a dream, think big, think bold". So it was a great challenge to take material that we've already cut for our mix and morph it on the mix stage to transform it. We'd set up a keyboard on the mix stage to be able to react to the changing situation. At points, we'd add new elements at the last minute, and it'd be like, "I'm just going to write some drums real quick". It was crazy but it worked.

NS   It's so well put together in my ears. The balance, the slick technique, everything is so finely balanced all of the time.

Depth, Perspective, and Workflows

NS   Perspective is another thing that I noted, the layers of distance. As an academic I'm writing on the topic of perception of acoustic space, and in your work it was so detailed.

I'm fascinated by the combination of the delays and reverbs that you use and how they created these zones that were interacting but were very clearly defined. You could hear all those different distances. That was very impressive.

OB   What you heard was our 7.1 down mix of our Atmos, which is pretty much your Atmos bed. My background session was probably 250 tracks wide of winds, crickets, etc., and everything was mostly mono. Sometimes, I'd be working with a whole bunch of mono sounds, but in places I would create a 5.1 bed.

The aim was to create this big depth of field, and you do this through creating very specific, discrete channels that only play in the backs, sides or fronts. Having that distinction, rather than sending it everywhere, will make your track feel more open and less mono-y.

NS   Yes, there's a lot of subtlety in how things are treated and even how they move. I think you're talking about very short delays, right?

OB   Yes, we use Slapper[61] a lot. You can add Slapper, and then you can choose your speaker channels that you want your delay to go to, and you can change the milliseconds, depending on your speaker configuration on your plug-in.

From there we have it going into a reverb, and then via EQ into another delay which is just going to the back speakers. So, your reverb has an extra delay on it, and using the EQ you can tune which frequency you want your delay to hit. There are probably four or five plug-ins just for delay in some instances.

NS   You've described a complex workflow, are you trying to position this sound in the narrative or position in a specific space?

OB   Because we need to do a lot of different things quickly, we need to be able to get to anything at any time. On the effects side, there were probably over 750 tracks of sound at once, split across two different machines. We probably had seven machines all hooked up together and two different recorders going, because we record massive amounts of stems at the same time. We record our Atmos, 7.1, 5.1 re-renders, our two track printmaster and all of our 5.1 and 7.1 stems (Dialogue, ADR, Group, FX, BG's, Design, Foley) all at the same time.

Everything's already set up in our templates, and you get very used to working in that way because you always have to be so fast [laughter]. For example, if my mix partner is working on mixing backgrounds and he needs some exterior slap, he already has in his chain of stems a room, a chamber, a long, a plate, delays. And then after all of those, he has more creative "have fun" plug-ins.

Dialogue reverb is a whole different thing. I always have mono reverb as well as many different sorts of stereo and surround reverbs, but I'll always separate out the returns into individual mono feeds, so that you have direct control over how much you want in the back. If you just have a multi-track reverb return you're really limited on how your track sounds. For instance, I love reverb on dialogue. I put it on everything. Whether it's a little bit or not, it helps just make the dialogue feel a little bit more embodied in the track. And with the mono reverb feeds, if they're off camera then you can easily raise your surrounds on reverb if you need to.

People think that the way I do stuff is complicated, but I don't want to tie my hands.

Narrative Clarity

I   *I wanted to ask about the intelligibility of the underlying concept. Is there something about the clarity and being able to stay true to that original idea or being clear on what the idea is, and not getting too drawn away from that? That ability to know where the boundaries are and the limits are, if that makes sense?*

OB Well, the story is king, right? The dialogue is king. The audience needs to know what's happening. If you're helping tell that story, create that mood, then you're doing a great job. With *Underground Railroad,* Barry [Jenkins] holds on shots, directs and cuts his story for sound and music, which really allows us as Sound Designers to take the opportunity to do something cool. And it's always changing.

But it all depends on the project and the story. For example, comedies are "b-b-b-b… joke", "d-d-d-d-d… joke". Let's have one more second between each joke, so the audience can laugh enough times to then hear the next joke. There's not even any space for room tone in those movies, it's all about the beats and the rhythm of the joke, which is interesting in itself.

Other times the story allows for more creativity.

NS Yes, I appreciated this. There were two or three moments in that episode where there was space, and then the sound took over. The train at the end. That was awesome, I remember my jaw dropping for that!

OB Yes, she goes down to the train track, and puts her ear to it, and the whole sound-scape changes just for that moment into what we hear in her ear canal.

Sonic Characters and Transitions

I   *Do you each think about sounds as characters? That they have agency, that they're an active part of the story, moving with their own behaviours?*

NS Everything that happens in the musical narrative is a result of an interaction. I want the listener to be able to identify a reason why things are happening, and to thus be able to recognise any tension that results from these interactions.

This is also something that I heard a lot in Onnalee's work. For example, whether you're streaming out layers because you have a cut approaching or you're thinning out the texture, the attack points are functioning to take things out or bring things in.

At least, that is how I heard it! Is that how you think about it?

OB I'm a big transition person. Every cut should have a meaning. And I almost want to get a tee-shirt that says "protect the cut".

Any big event is a way to get out of something else without the audience knowing. Let me put something in to catch their ear and draw attention to something else,
and then they won't notice that I'm subtly transforming this other thing. Whatever
you can do to make it work smoothly.

NS   Yes, yes!

OB   It all depends on what story is told, whether it's nice to have a hard cut, or stuff that
starts really early and morphs into something that foreshadows the cut, and creates
a sense of mystery and pull. I was obsessed with watching Coen Brothers[62] movies
as their transitions and the sound design that Skip Lievsay[63] does are really great.

When I started working on *Game of Thrones*,[64] the Producer would have us do a
chair only pass, "Okay everyone, we're just going to watch this episode and we're
just going to look at chairs". And so, every time any character would move in a chair
you'd have to have a reason why that character is moving and the chair itself needed
to have its own unique character. It really made me start thinking about all movies
in that way. And it helped me to think in great detail, "Why are they moving their
hand? And what are they doing there with that? Are they uncomfortable? Are they
nervous?" So, I almost think of each character as a sound particle or a sound unit.

In the last episode of *Underground Railroad*, which takes place on this planta-
tion, the main character almost loses her mind because there's so much negativity.
To reflect this sonically I was articulating frequency ranges, transforming cicadas to
go "ch-ch-ch-cc-c zing", and almost create these very subconscious high tones as
the episode progresses. And then shock the audience by making another section
completely silent.

Abstraction and Reality – Hyper-Real

I   *I wonder whether we could talk a bit about this question of abstraction versus reality? The playing with and shifting the sense of reality, do you feel constrained by it?
Or how far can you push it?*

OB   In *Underground Railroad* episode one, the temp track that the Picture Editors used
to develop the cut for the opening sequence, had the baby blaring through the
whole thing, and they kept saying, "we've absolutely got to have the baby present,
it's a big story point".

When we got to the mix stage, I said, "I'm telling you that right now, we're not
having that baby all the way through". Why make it so on the nose on what the audience is supposed to feel? Why can't they figure out what they're supposed to feel on
their own? The baby cry can come and go, so my ear can reintroduce it each time.

Just like in ballet, if you keep doing the same combination over and over again,
the audience will want to look at their phone or shift in their seat. So how can you
change up five steps here or there, but still have the continuity to keep the audience
engaged with what's going on? If you keep hearing the same frequency tone, your
ear will tune it out. So how can you keep it moving, but have it there the whole
time?

It's the same thing with noisy dialogue scenes. I feel that it is a huge mistake to
strip out the broadband noise behind dialogue with iZotope,[65] as it takes a lot of the
life out of what's recorded on set. If your track is noisy and you have people talking,
just have the noise start at the beginning of the cut, and in about ten seconds the

audience won't even hear the noise anymore, they'll just focus on the words. Then use slight iZotope on the dialogue if you need.

If I want the audience to continue to listen to the track, then I need to change it up. I am trying to push the envelope and make a regular scene, even a talky scene, sound weird or different.

For example, in the episode that you watched, the mean slave catcher is a character called Ridgeway, played by an amazing actor Joel Edgerton, and any time he meets the main character (who is the daughter of a slave that he's trying to catch), and whenever he looks on screen directly into the camera, his dialogue is in all of the channels. But if he's off screen, it's not. Even just this subtle spread of the dialogue makes the audience feel, "wow, something's different about that character!". They can't really put their finger on why, but they really notice it when it goes away because it makes it feel really intimate all of a sudden, so it can make a character feel really scary.

I   *I think it's the intensity of that sequence you shared that's the most impressive thing. It grabs you right at the beginning of this episode, and it just carries you through. There's almost no time to breathe as the audience; it's as if you're there. And maintaining that intensity through the duration so that it still feels intense. Like you were saying about the baby, if the baby had cried all of the way through the scene, then it would also have lost a bit of its impact. When it moves away and it shifts and changes and comes back, then it still retains that impact because something's changed about it.*

OB   Exactly, yes. You care about it still. Instead of thinking "Ah it's so annoying, why is this baby wailing in my ear"? [laughter].

Notes

1 A special thanks to Pro Sound Effects for sharing excerpt sound recordings from the *Sound Mountain Archive* of Ann Kroeber and Alan Splet.
2 Westerkamp, H. (1992). A Walk Through the City. *Transformations*. [CD]. Canada: Empreintes DIGITALes.
3 Westerkamp, H. (2002). Attending to Sacred Matters. *Into India* [CD]. Canada: earsay music.
4 Furley, S. (2011). *Ann Kroeber Special: A Pioneering Sound Woman*. Available at: https://designingsound.org/2011/10/07/ann-kroeber-special-a-pioneering-sound-woman/.
5 Schafer, R. M. (1994). *The Soundscape: Our Sonic Environment and the Tuning of the World*. Rochester: Destiny Books.
6 David Lynch (b.1946) is an American Filmmaker.
7 Martin Stig Andersen (b.1973) is a Danish Composer and Sound Designer.
8 *Wolfenstein II: The New Colossus*. (2017). Microsoft Windows, PlayStation 4, Xbox One and Nintendo Switch. Bethesda: Bethesda Softworks.
9 Gus Van Sant (b.1952) is an American Film Director, Producer, Photographer and Musician.
10 *Elephant*. (2003). Directed by Gus Van Sant. [Feature film]. Burbank, CA: Fine Line Features, & New York: HBO Films.
11 Westerkamp, H. (1992b). Beneath the Forest Floor. *Transformations*. [CD]. Canada: Empreintes DIGITALes.
12 Westerkamp, H. (1999). Moments of Laughter. *Breaking News*. [CD]. Canada: earsay music.
13 A desert patch near the Bolsón de Mapimí in Durango, Mexico.
14 Westerkamp, H. (1987). Cricket Voice. *Transformations*. [CD]. Canada: Empreintes DIGITALes.
15 Nagra is a brand of high-quality portable audio recording devices, which captured sounds on reel-to-reel 1/4" magnetic tape.
16 *Dune*. (1984). Directed by David Lynch. [Feature film]. US: Universal Pictures.

17 Wishart, T. (2000). Two Women. *Voiceprints*. [CD]. New York: EMF Media.

18 Wishart, T. (2014b). Encounters in the Republic of Heaven. *30 Jahre Inventionen VII 1982– 2012*. [CD]. Germany: Edition RZ.

19 Wishart, T. (2007). *Fabulous Paris: A Virtual Oratorio*. Available at: https://www.discogs.com/ release/1492681-Trevor-Wishart-Fabulous-Paris-A-Virtual-Oratorio. (Accessed: 3 November 2022).

20 Wishart, T. (2020). The Garden of Earthly Delights. *The Garden of Earthly Delights*. [CD]. London: ICR Distribution.

21 *Mowgli: Legend of the Jungle*. (2018). Directed by Andy Serkis. [Film]. California, US: Netflix, Inc.

22 Gravity. (2013). Directed by Alfonso Cuarón. [Film]. California, US: Warner Bros. Pictures.

23 Wishart, T. (1978). Red Bird. *Red Bird (A Political Prisoner's Dream)* [Vinyl]. York: York Electronic Studios.

24 Alfonso Cuarón Orozco (b.1961) is a Mexican Filmmaker.

25 *Bohemian Rhapsody*. (2018). Directed by Bryan Singer. [Film] US: 20th Century Fox.

26 Vlad, G. (2020). *Intense Dusk Chorus In The Borneo Rainforest*. Available at: https://soundcloud. com/georgevlad/dusk-borneo (Accessed: 1 August 2021).

27 KMRU. (2020). Continual. *Continual*. [EP]. Available at: https://kmru.bandcamp.com/album/ continual (Accessed: 1 August 2021).

28 An example recording by George Vlad can be heard here: https://soundcloud.com/georgevlad/ amazon-larvae-feast?in=georgevlad/sets/amazon-rainforest.

29 Vlad, G. (2017a). *2017 Sound recording expedition to Senegal*. Available at: https://youtu.be/ C4sLlLWjDf0 (Accessed: 1 August 2021).

30 Chris Watson (b.1952) is a Wildlife Sound Recordist - https://chriswatson.net/.

31 Kibera is the largest informal settlement (slum) in Africa. Situated 4.1 miles from the centre of Nairobi, it is estimated to be home to 250,000 people.

32 Vlad, G. (2017b). Field recording trip to South Africa – 2016. Available at: https://mindful-audio.com/blog/2017/9/4/field-recording-trip-to-south-africa-2016 (Accessed: 1 August 2021).

33 Ambisonics is a format of surround sound recording.

34 Omnidirectional microphones capture sounds coming from all directions onto one channel.

35 Vlad, G. (2021a). *Chromatic rain in the Namib desert - Nature sounds*. Available at: https:// youtu.be/P09ZwPBwenM (Accessed: 1 August 2021).

36 A microphone which is designed to be used underwater for recording and listening.

37 Mahtani, A. (2015). *Inversions*. Available at: https://soundcloud.com/annie-mahtani/inversions-2015-stereo-mixdown (Accessed: 3 November 2022).

38 Mahtani, A. (2019c). Racines Tordues. *Racines*. [CD]. Canada: Empreintes DIGITALes.

39 Mahtani, A. (2019d). 'Round Midnight. *Racines*. [CD]. Canada: Empreintes DIGITALes.

40 *The Little Stranger*. (2018). Directed by Lenny Abrahamson. [Film]. France: Pathé Distribution and UK and Ireland: 20th Century Fox.

41 Leonard Ian Abrahamson (b.1966), an Irish Film and Television Director: https://www.imdb. com/name/nm1049433/.

42 Nathan Nugent is an Irish Film Editor known for working with Director Lenny Abrahamson: https://www.imdb.com/name/nm1524435/.

43 Domhnall Gleeson (b.1983) is an Irish Actor, Director and Screenwriter.

44 Stephen Rennicks (b.1972) is an Irish Musician and Film Score Composer based in Dublin.

45 Also known as a piezo microphone, is a form of microphone that senses audio vibrations through contact with solid objects.

46 A microphone which converts the sounds of the human body and/or objects into electrical signals, which allow amplification, processing, and recording of the sounds.

47 The sounds of a given location or space.

48 A large town in County Fermanagh, Northern Ireland.

49 Electroacoustic Music has a rich history of loudspeaker concert performance which stretches back to the earliest experiments in recorded sound. *Diffusion* is the name given to the performance of articulating sounds spatially in real time. Annie Mahtani is the current Co-Director of BEAST (Birmingham Electroacoustic Sound Theatre) BEAST, a Loudspeaker Orchestra performance system. See www.beast.bham.ac.uk.

50 A live distribution of sound throughout a space using multiple loudspeakers, and often typically heard within concert hall settings.
51 The aural equivalent of a point-of-view shot.
52 Formed in 2014, a Research Institute to provide science, social science and cultural research to global forested landscapes.
53 A high sensitivity lavalier microphones.
54 Stavropoulos, N. (2019). *Claustro*. Available at: https://soundcloud.com/nikos-stavropoulos/claustro-stereo-reduction (Accessed: 3 November 2022).
55 *The Shining*. (1980). Directed by Stanley Kubrick. US, Warner Bros. & UK, Columbia-EMI-Warner Distributors.
56 Further details published: Stavropoulos, N. (2018). *Inside the intimate zone: The case of aural micro-space in multichannel compositional practice*. In: Proceedings of the 15th Sound and Music Computing Conference (SMC2018). Sound and Music Computing Network, pp.113–117.
57 Music creation software developed by Ina-GRM.
58 *The Underground Railroad*. (2021). Directed by Barry Jenkins [Drama Series]. US, Amazon Prime Video.
59 Jonty Harrison (b.1952) is an Electroacoustic Music Composer based in the UK.
60 Barry Jenkins (b.1979) is an American Filmmaker.
61 Slapper is a multitap delay plugin by Cargo Cult – https://www.thecargocult.nz/products/slapper.
62 Joel Daniel Coen (b.1954) and Ethan Jesse Coen (b.1957) are American Filmmakers.
63 Skip Lievsay is a Supervising Sound Editor, Re-Recording Mixer and Sound Designer for film and television, based in the US.
64 *Game of Thrones*. (2011–2019). [TV Series]. US, Warner Bros. Television Distribution.
65 Audio software and plug-ins for mixing, mastering and restoration of sounds.

6 Celebrating the Art of Sound

Within this book we set out to discover what really constitutes creativity within sonic practice, exploring the drivers behind the decision-making processes employed by world leading professionals. The ultimate goal has been to find if we could identify new and better ways to talk about (and therefore think about) sound practices.

In the introduction we defined a series of interrelated ambitions that helped us to structure our investigations. We wanted to:

- *Reveal new insights through attention to the creative, communicative, and aesthetic processes of making.*
- *Encourage a greater appreciation for the artistry of Film Sound practices.*
- *Foster dialogues between Film Sound and Electroacoustic Music which help to challenge the traditional ways in which we think about these respective creative practices.*
- *Use these new insights and dialogues to further widen the diversity of creative professionals engaged in Electroacoustic Music and Film Sound.*

Through diverse conversations with expert practitioners, we have documented and recognised the complicated, and sometimes messy details of creative practice. Our results have demonstrated that creativity in sound (whether in Film Sound or Electroacoustic Music) does not unfold in a direct linear fashion (e.g. Pre-formed Idea → Action → Result), but via much more complex iterative interrelationships of inspiration, material, and action Creativity is not a direct and 'top down' process but an emergent property of people shaping materials within iterative flows and processes. If we want to truly understand creativity in sound practice, then we must pay attention to how these flows and processes operate.

Key to this understanding is the need to move beyond a direct descriptive focus on actions taken (*what was done*), to access and foreground the rationales and reasons behind *why* actions might be taken. The implication of such a shift is to recognise that we must redirect our attention away from a narrow focus on the individual finished result, towards the *process* of creation itself. It is a fallacy to seek 'the one correct solution' in creative practice, instead we must recognise and celebrate the many possible outcomes that might emerge and the rich complexity of decision-making processes in action. We should not seek to identify universal answers, but embrace the multiplicity and possibility of the many potential answers that might exist.

Because we were able to speak with so many creative individuals within one structured research project, it was possible to review a diversity of approaches simultaneously.

DOI: 10.4324/9781003163077-6

This enabled us to demonstrate new ways in which we might try to understand creative practice in sound, shifting away from a universal approach – focusing on absolute facts, secret formulas and 'correct' ways of doing things – towards one that embraces a diversity of individual perspectives, and celebrates this multiplicity as the root of creativity across both Film Sound and Electroacoustic Music. The chorus of voices within this book provided access to deeper underlying insights through a range of unique approaches to creative sonic practice, helping us to move away from a focus on individual works, specific projects, or unique personal approaches. Diversity of perspective has therefore provided understandings of creativity that would not have been accessible without bringing together so many individual perspectives in one place.

It is a simple shift – from absolute notions to relative possibilities – that underlies all of our findings. As we will expand below, the most significant ramification of this is to highlight and celebrate the roles of exploration, innovation and discovery within sonic practices, and how these situate both Electroacoustic Music and Film Sound as fundamentally artistic acts of human communication. In this final chapter we will demonstrate the positive implications of this shift in framing creativity as an unfolding process, and discuss some concrete examples which reveal how this change in thinking might be applied to benefit our fields.

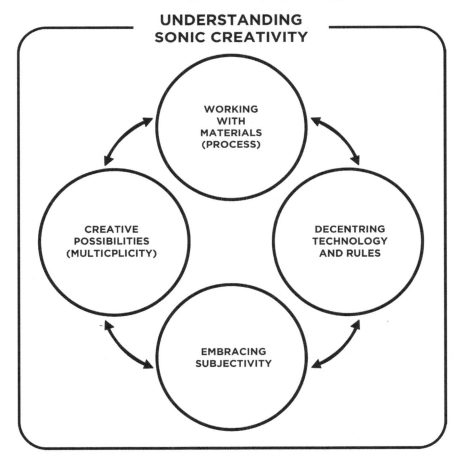

Diagram Three The interrelated components of our understanding of sonic creativity.

6.1 Creative Possibilities (Multiplicity)

Our conversations revealed how the secret of 'successful' creative practice is not the discovery of 'the one' absolute solution, but a negotiation of the many possibilities available. There is never a single 'correct' solution to a creative challenge, there are always many potential solutions and this multiplicity is inherent to work in sound.

Our outputs (whatever form they might take) all emerge through a process of careful and repeated creative engagement, in response to the complex associations that exist between:

- Creative intention.
- Imagined audiences.
- Collaborative discussion.
- Evolving properties of the work's materials (sonic or visual).
- With all above modulated by unexpected and chance encounters.

As we create we are working *with* materials, engaging them in dialogue, constantly reflecting on the impact that each decision made has upon the balance of the overall result. Many small choices gradually build up to shape the final outcome, with repeated reflection upon the evolving materials guiding the process. At any point in the creative process, there are multiple possible choices and multiple streams of input informing the decision that might be taken. The points across 'Our Map of Sonic Creativity' reflect some of the decision-making factors which might be employed by creatives (Diagram 2, p.28)[1]. But these factors should not be considered as absolute rules to be followed, only as trends in possible approaches.

If we mask this inherrent multiplicity of the creative process, we can end up imagining that there was only ever one possible outcome to the task at hand. We then risk becoming fixated on trying to find that one true 'correct' sound, instead of acknowledging the range of 'possible' sounds that might work. This is an example of universalisation in practice. Too often, studies in Film Sound and Electroacoustic Music focus upon final artefacts, and while it might seem logical to attend only to the final created artefact on the surface (after all, we are commissioned to deliver outputs), this can have complex knock-on effects for how we imagine the creative process. Focusing only on the final result creates a false impression of creative practice by introducing clear separation between the artefact created, and the world around it. By ignoring the inherent multiplicity of the creative process, we mask the fact that the final artefact is only one possible outcome of a wide range of potential possibilities.[2] And as such, the final output is disconnected from the messy reality of its production (and its perception by audiences). The artefact is imagined to be an almost timeless entity, something which sprang into existence in the way that it was always going to be. In so doing, the many opportunities, tribulations, and choices of the creative process are collapsed and replaced with the simplistic idea that what emerges at the end was exactly what was intended all along. When this happens, an imagined 'underlying conceptual idea' becomes centred; while the materials themselves, the processes of making, and the final audience, are all denied agency. The artefact is imagined to be a direct realisation of a pre-formed conceptual idea. In such a situation we end up with a fragmentary and skewed perspective of creativity. In much the same way that Susan Langer describes the division between symbols and phenomena, "in talking about things we have conceptions of them, not the things in themselves" (Langer 1957, p.60) we can easily become mistaken in thinking of sounds as absolute objects, fixed markers of communication with inherent meaning.

In recognising that there are many possible solutions – building out from underlying ethos' of sonic practice, via applied approaches to more concrete techniques that have directly tangible relationship to materials – we can celebrate multifaceted, unique and individual processes of decision-making within sound practice.

6.1.1 Art, Craft, Process

This conceptual separation of the final creative artefact away from the messy material process of its creation also acts to perpetuate a division between the notion of creative 'idea' and that of the technical 'realisation' of the work. This has political ramifications, constructing a hierarchical division between artistic creative inspiration and the 'craft' process of shaping of matter. In such a formulation, the abstract idea is valued more highly than the craft of its technical realisation. Such a division between idea and matter has many deeply embedded cultural associations, not least in notions of the divine versus the worldly, and by envisioning creative practice as a rendering of a pre-formed creative vision, art takes on characteristics of the divine with the impression that creative vision exists in a purely conceptual form, something intangible. The myth of the omnipotent creator, acting from above and in total control over the decision-making process, provides a comforting air of certainty in the face of the real complexity and messiness of the creation process. It is much easier to imagine that art unfolds from a complete conceptual idea pre-formed at the very start, and that the very best ideas are reserved to those special few with unique insight and access to the most 'true' and 'correct' answers; those gifted with 'genius'.

However, if this were true then world leading professionals might logically be considered more likely to possess the traits of the 'genius' artist, with the presence of genius in their person a contributing factor in their world leading status. As such, the status and esteem by which many contributors in our study are held should have ensured that our results would demonstrate clear examples of their 'genius' in action, including innate access to the 'correct' sonic solutions and entirely pre-formed conceptual creative ideas at the very start of their creative journey. However, the generous and honest reflections in this volume have demonstrated that these assumptions of omniscient genius are a fallacy. If those world leading experts, who are most likely to fulfil the conditions of 'genius', reveal that their practice is one of multiplicity, discovery, and creative possibility, then we must necessarily reject any notion of universal creative solutions as singular 'correct' responses. Creativity is not the result of access to divine ideas, but of hard work and repeated engagement with material in unfolding process.

This requires a shift in how we celebrate the creative process, widening our attention from *'creation as realisation of a singular idea'* to *'creation as the negotiation of possibilities'*. It is a movement from absolutes to potentials, which demands that we engage the inherent agency of materials and the 'other' – an imagined listener – as essential factors in the creative process.

6.2 Working *With* Materials (Process)

Our conversations repeatedly revealed the creative process to be one of unfolding, growth, and development. Sound practitioners do not impose ideas upon an external world of inert 'raw' materials, instead they *join forces with* their materials to discover what emerges.

> Far from standing aloof, imposing [their] designs on a world that is ready and wait-ing to receive them, [... they] intervene in worldly processes that are already going on, [...] adding [their] own impetus to the forces and energies in play.
>
> (Ingold 2013, p.21)

This is a non-hierarchical process in which sounds have as much power to influence the creative as the creative has to influence them. It is a shift from thinking about creative practice as a top down process of *working on* materials, to one of *working with* materials, an active dialogue, and conversation (not a hierarchical dictation). Such a shift, to ap-preciate making as an active process of *unfolding with* materials, has radical implications for how we consider, value, and recognise creative practice and contribution in sound.

To recognise 'process' is to celebrate the creative journey which lies at the heart of making. To appreciate that excellence in sonic practice is not constituted by direct reali-sation of a fantastic pre-meditated idea, but through sustained engagement with the rich possibilities in sound. Such negotiation of possibilities must embrace the fact that the unique contingencies of each context and different combinations of inputs might lead to varied outputs. Give the same brief to different creatives and the result will reflect a range of different possible outcomes. Each creative will make their own material choices and – listening closely to how the materials themselves feed back to them – will shape and develop their inspirations along with the material, unfolding both together. This interre-lationship of inspiration, material action, and reflection, unites creative thought and craft practice. There is no hierarchical division. All are linked equally within a chain of unfold-ing; a series of loops which gradually modulate forward towards an end that emerges from the noise of many possible outcomes (Diagram 4). Material and action are so intimately connected as to be indivisible, an Assemblage (*Principles*).

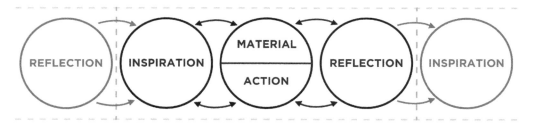

Diagram Four Iterative cycles of inspiration, material action and reflection in creative practice.

Inspirations are not universal overarching ideas, they are situated in the moment, de-fined by the current context and the balance of the materials at one specific point in time. They are active intentions, *ideas for action* which shift and modulate as the development of the work progresses. *Reflection* feeds back into new *inspiration*, creating the next loop in the chain, responding to the characteristics of the material as perceived after each action has been realised. Thus, these inspirations might be described as situated, both in terms of the perspective of the individual creative artist, and by the unfolding con-text of the work and its materials from moment to moment. This notion of the 'situated idea' is drawn from Haraway's concept of situated knowledges, which are the individual perspectives from which each individual engages with the world. They are embodied positions that emerge from located subjects (views from somewhere rather than views from nowhere), are necessarily partial perspectives (as each individual has a unique position framed by their own unique bounds), and are geographically and historically specific (continually structuring and being structured by social conditions) (Holloway 2011, p.222). The relationship of creator to the shifting balance of options and opportunities within making is necessarily a situated and partial perspective. It is relative, defined by the variables of the moment and the partial knowledge of other potential opportunities connected in their unfolding movement.[3] Creative action is about possibility, negotiating the balance of options in? concept? with the evolution of unfolding networks of inspiration, material, action, and reflection.

Within each moment, it is not only the creator that has agency, the material properties too have agency, it is for this reason that we imagine action and material to be indivisible in our Diagram 4 above. As such, we extend this notion of a situated partial position to every choice that an individual makes, providing an avenue for understanding creative process at the moment of its unfolding. As Haraway states,

> Situated knowledges require that the object of knowledge be pictured as an actor and agent, not a screen or a ground or a resource, never [...] as a slave to the master that closes off the dialectic in his unique agency and authorship of 'objective' knowledge.
>
> (Haraway 1991, p.198)

This is a reaffirmation of Ingold's argument of collaboration between artist and material, both united in process and a rejection of 'universal' knowledges (such as those suggested by the notion of the genius). Creative inquiry necessarily comes from a partial position, it is a problem-solving action seeking an outcome and resolution. The unknown factor is the very thing that compels action, a lack of certainty that dives us forward. If the endpoint were known, then what impetus for the imagination? There would be no discovery, and no possibility for chance acts or creative innovation en-route. Creative practice is a Process of Discovery *(Principles)*, a journey of coming to know what the outcome might be, revealed not from overarching creative vision, but from continued engagement with the materials of the work. Practitioners engage in a Constant Enquiry *(Principles)* working with the affordances of materials in new contexts to deliver the best possible outcomes.

The art of sound, is engendered in action with sound materials, there is no division between sound art and sound craft. Creativity is constantly enacted within the crafting and articulation of sound, working through inspiration, in action with materials, pushing to explore the available Plurality of Potential *(Principles)*.

6.3 Embracing Subjectivity

Sounds do not mean 'one' thing, they have the potential to mean many things, and this feature is fundamental to their communicative potential. As noted elsewhere,

> It is this property that allows Foley artists to perform the 'real-world' diegetic sounds for a given film in a recording studio context and yet have the film audience attribute these sounds to the actions they see on the screen.
>
> (Hill 2017)

Listeners do not simply receive transmitted information, but actively participate in the construction of its meaning through their perception.[4] Subjectivity – the influence of personal beliefs and feelings – affords that one sound can have many different meanings depending on the lived experience of the listener perceiving it. This is a fundamental challenge to the universalist viewpoint which considers there to be fixed truths which are unchanged by shifting contexts.

This is not to abandon ourselves to unlimited fluid intersubjectivity, a 'free-for-all' where everything or anything might be argued as standing in for or meaning anything else (and any creative decision justified as acceptable). It is rather a "prioritization of affective experience over the conceptual" (Knight-Hill 2020a, p.61) a concession that objective positions are necessarily partial. Objectivity is not denied, but it is denied universality; and this rejection of universal truth frees us to a more accurate, realistic, and nuanced understanding of the creative act.[5]

To accept subjectivity is not to deny creative intention. Rather, it is simply to shift our understanding of creative practice from one of message transmission, to that of communication, dialogue, and exchange. A universalist viewpoint might assume that, if the artist cannot ensure the direct transmission of an idea (due to audience agency), then this artist is somehow lacking in the proficiency to articulate their creative vision. But as Haraway states,

> The topography of subjectivity is multi-dimensional […]. The knowing self is partial in all its guises, never finished, whole, simply there and original; it is always constructed and stitched together imperfectly, and therefore able to join with another, to see together without claiming to be another.
>
> (Haraway 1991, p.193)

Thus, the subjective listener is an active listener, engaged in the process. They are drawn in to become part of the storytelling process, active participants in a shared experience, completing it by their own act of listening. That the audience are required participants for communication to succeed is therefore a vitally important consideration for those working in creative practice to seriously engage with.

If we want to communicate meaningfully, we must do so in a way that connects with people's experiences. The fact that a listener can project their own experience into the interpretation of a sound is core to the aesthetic experience of listening. As such, we must find ways to embrace multiple individual subject-positions in our work, accepting that we can never fully know the other, but that we can build connection by seeking to access shared features of underlying experience.[6] Communication is always reliant on negotiation and a shared willingness to find meaning. Messiness and the potential for misunderstanding are inherent features within all forms of human communication. And, rather than see this openness as a negative (as dominant universalist communication models might suggest), we can instead flip the dynamic and celebrate this richness of variability as the very foundation of what makes experiencesvaluable.[6] The embrace of subjectivity is thus an acceptance that there are no absolute signifiers, no universal signals, no one single 'correct' approach to a creative challenge. Experiences are meaningful because they are meaningful to us. Subjectivity values every listener's individual perspective and welcomes them into the perceptual process as an active participant.

Creative outputs emerge through a complex web of connections that draw upon affective and empathic potentials in communication. In turn, they construct and subvert representations of reality and expectation, and, in so doing, they celebrate the rich potentials of meaning, connection, and possibility.

As we developed our analysis of the conversations in this book, we identified again and again, the recurrence of practices engaging with subjectivity and communication in a multitude of formulations and framings. These became repeated features across Our Map of Sonic Creativity, embedded within underlying *Principles*, at the core of many *Approaches*, and directly applied via *Techniques*:

- Emotion/Feeling/Affect *(Principles)*
- Sonic Communication *(Principles)*
- Plurality of Potential *(Principles)*
- Authenticity *(Approaches)*
- Sonic Storytelling *(Approaches)*
- Sound & Character *(Approaches)*
- Openness *(Approaches)*
- Reality *(Techniques)*
- Listening Empathy *(Techniques)*

When highlighted across our map we can observe the spread and interconnectivity of these features across all sectors of creative experience, demonstrating subjectivity to be a significant concern across a wide range of decision-making processes (Diagram 5).

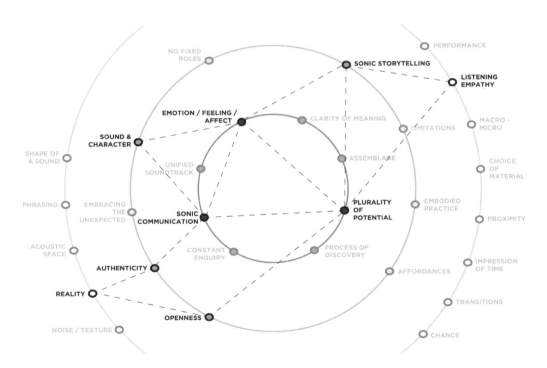

Diagram Five Annotated Map of Sonic Creativity, highlighting practices of communication, dialogue and exchange.

This ubiquitous presence demonstrates that the fundamental driver of sonic practice is not centred in tools, techniques, or even sound itself, but in people, and the ways in which those people relate to, communicate, and find meaning within sound. Artistry within sound practice is constituted by working to shape contexts and relationships between materials in such a way that will carry the attention of the listener to respond to facets of sonic (or visual) materials which convey shared feelings and understandings. Recognising that subjectivity is not an uncontrolled randomness of uncertainty but rather a complex matrix of possibility, allows us to affirmatively embrace its features and their ability to more accurately reflect the complex facets of the creative process. Genuine and deep acceptance of the role of the listener demands a fundamental shift in the way we frame our thinking around sonic practices, moving away from the technological to the human.

6.4 De-Centring Technology and Rules

Sound practices are underpinned by technical skills and proficiency with tools, but it is the creative direction of these within an unfolding Process of Discovery (Principles) that is the key factor in delivering effective results. It is common for discourses around both Film Sound and Electroacoustic Music practice to centre tools, technology, and practical steps of craft via direct descriptions of techniques applied and steps taken to edit and transform sound. An example of such a description might be:

> "I routed my audio via an external bus to an analogue outboard EQ with a low cut at 60Hz, a boost at around 3000Hz, and then re-digitised it via a pre-amp which added distortion for increased texture and warmth".

On the surface, such descriptions purport to reveal something about creative practice, but in actuality they provide only a cursory summary of actions taken, revealing nothing about the underlying process of creative decision-making that inform *why* such processes were enacted.

A descriptive discourse marginalises the human element in favour of a focus on artefacts as fixed entities, sound objects. When we focus on sound objects we risk conceptualising them as absolute separate entities, clearly demarcated in terms of space but continuous in terms of time, imagining that they always existed. A focus on fixed sound objects draws our attention away from the dynamic flux of process, and situates materials as inert and 'raw' before they are transformed by a human mind into a 'cultural object' with significance. Ingold describes this as imagining practice as a 'project':

> [We] start with an idea in mind, of what we want to achieve, and with a supply of the raw material needed to achieve it. And it is to finish at the moment when the material has taken on the intended form.
>
> (Ingold 2013, p.19)

This is an inherently hierarchical way to envisage practice, in which the creative is situated omnipotently above their materials, enacting decisions upon them. This is a return to the separation between material and idea, discussed above. If the reality of creative practice was truly that of a 'project', in which the practitioner applies their enlightened 'genius' vision upon materials, there would be no space for chance encounters or the unexpected: no happy accidents, no moments of discovery, no purpose for experimentation, or the development of new ideas and approaches.[7]

The simplified notion of the 'project' with its fixed and (apparently) absolute outcomes, is a dangerous surrogate for the reality of creative process. Its universalism pushes us towards an analytical categorisation of sound which artificially segments and splits the sound object from its materiality (e.g. "Is this sound diegetic?", "Is this music or sound?", "Is this real or not?") denying the richness of situated perspectives and the possibility of openness[8]. Universalism implies single 'one-size-fits-all' solutions and can be used to push discussions of creativity away from their complex realities towards denying the role of material, the multiplicity of possible solutions and the active contribution of the listening audience.

Such simplification frames making as an exercise in purely technical realisation. As such, descriptive discourses actively mask our understanding of sonic creativity because they:

- Centre on the one solution enacted (and not the many possible options available).
- Fixate attention on the final finished artefact (masking iterative processes of creative development).
- Risk implying there is one singular and absolute solution to a creative challenge (empowering tool and technique over creative sensibility).
- Reduce or deny the active role of the listener in perceiving the result.
- Conceptually position the creative practitioner over and above sound materials in a hierarchical power dynamic (working *on* sounds, instead of working *with* them).

In stark contrast to these definitions, our interviews demonstrate the rich multiplicity and diversity of creative possibility. Creative practitioners value and embrace the unexpected in creative processes of discovery (Process of Discovery, Constant Enquiry (*Principles*); Embracing the Unexpected (*Approaches*); Chance (*Techniques*)). Making is a complex and iterative process and while practitioners might begin with an intention to guide their creative actions, they have no overarching plan of the 'perfect' form to be realised. Instead, they positively embrace chance and respond to the agency of materials as they evolve within the creative process.

That the final outcome is unknown is at the very heart of the creative challenge. This is not to abandon ourselves to unlimited variation, but to acknowledge that our focus needs to be shifted away from universal and declarative approaches to describing and understanding practice. That the unexpected can and does perform such a significant function within the creative process, demonstrates that 'pure' pre-conceived creative visions are a fallacy. Everything is contingent. There are no absolute fixed rules, and it is not the fixed absolute artistic vision that is the source of sonic creativity, but people working through their inspirations.

This procession of ideas, leads us back around into Creative Possibilities (Multiplicity), and again through Working *with* Materials (Process), to Embracing Subjectivity, and a further De-Centring of Technology and absolute fixed rules. These interrelated factors are vital components in understanding the operation of sonic practices. They are mutually contingent and while perhaps not revolutionary in isolation, recognising their interrelated operation enables us to paint a picture of sonic practice which celebrates the situated creative possibilities available. It is a non-heiararchical mapping of process which demonstrates the rich multiplicity and creativity of sonic practice, and the roots of its powrful potential for artistic communication. These points also suggest calls to action in how we might further strengthen our field by having the confidence to embrace this knowledge and grow together.

6.5 What Does All This Mean in Practice?

Any disconnection between the ways in which we think about doing, and the ways in which we actually do, limits our ability to imagine, understand practice, and embrace the full creative possibilities available to us. As we have begun to realise, conventional simplified ideas about how creative practice operates, clash with our tangible understandings of doing in practice, as demonstrated by the results outlined by our study. Thus, it is important for us to work through the arguments above and to demonstrate their implications in the applied environments of work and education.. In so doing, we are not seeking to establish new sets of formal 'how to' rules, but to rather celebrate the rich complexity of potential approaches available.

The primary benefits of this exploration and discussion of sonic practices are that we:

6.5.1 Enable More Open and Honest Exchange – Sidestepping Gatekeepers

In recognising that making is an iterative unfolding discovery process guided by situated knowledge, we can liberate ourselves from a protective hoarding of 'trade secrets' or universal 'one size fits all' knowledge, because we accept that there are no direct secret 'answers' to success (beyond listening to others and continually experimenting and refining our approach).

Celebrating the journey of discovery, and demonstrating the complex and networked associations that operate within sonic creativity (including the significance of the unexpected and unknown) we can sidestep gatekeepers who might otherwise seek to control access to or validation of knowledge. We can act to empower new practitioners by breaking down the façade of perceived rule systems and the limitations they construct.[9]

In de-mystifying creative processes we can become more open and honest in how we share and discuss approaches to practice, and more open to new ideas, recognising that there are many possible approaches available..

6.5.2 Encourage Greater Confidence – Accepting Experimentation, Failure and Discovery

If we envisage the creative process as an unfolding Process of Discovery *(Principles)* we acknowledge that there are no limited 'correct answers' but many possibilities. As such, we need not limit ourselves to replicating common tropes or conventions. Instead, we can be more inventive and celebrate the manifold ways in which sounds *might* communicate.

When we mask our failures and processes of experimentation we risk perpetuating a false impression of the creative process. There becomes a disconnect between how creativity is conceived externally and how the practitioner knows the process to be. The result is a private acknowledgement of creativity as replete with failure and experimentation, but a public façade of creativity as the infallible rendering of clear assertive intention, and the enacting of pre-formed creative visions.

The significance of experimentation and failure can no longer be a solely private acknowledgement, but must form the basis of the very way in which we fundamentally conceptualise our practices. This is to flip the dominant mode of discussion from that of rules and 'how to' procedures, over to those of possibility and potential.

Being more freely inspired by the richness of potential, promotes an increased confidence in experimentation and creativity, embracing the complex possibilities of subjectivity, potential, and openness. As such, we can be more open to new ways of doing and less held to account by 'rules' or 'conventions'.

6.5.3 *Promote the Centring of the Human, Over the Technical*

Working directly with sound creates a very close relationship between material and practitioner. We become engrossed in our sounds and can begin to think of them as absolute objects with implicit meaning. We can come to think that everyone will hear them in the same way we do, with all of the attendant associations we have as creatives.[10] We begin to imagine our sounds as fixed artefacts, creating a false impression of universality within the way we imagine our sounds being perceived.

This shift towards universalisation denies the role of the listener and can lead us to overinflate the significance of tools and technical functions. We begin to focus on process and equipment, believing that they alone are able to impart powerful value into sounds.[11] We shift from working *with* materials – delicately shaping or iteratively experimenting – towards imagining that we work on materials, rendering fixed and inert artefacts. Pushing into this simplified space comes the saviour rhetoric in marketing of new tools and technology, framed as "enablers of aspiration" and acting as "catalyst[s] for imaginative possibility or economic opportunity" (Sporton 2015, p.60). Companies to position their latest tools and technologies as saviours, purporting to further extend the power of the artist offering a slice of omnipotence (they promise access to something of the divine genius). Tools are presented as offering an opportunity to overcome the limits of your human body and extend your potential, delimiting the human by providing something beyond what nature can provide alone (Sporton 2015, p.15). In the process they seek to both construct and reinforce the fantasty that simple and direct fixes to creative challenges do exist (if only you purchase their product).

Through our conversations and discussions we have demonstrated that the reality of sonic practices are far richer and more complex. We need to break out of the aloof creative mindset that technology companies like to promote, and, project ourselves outside of the studio or laboratory and into the concert hall or theatre. We need to remember that our audiences are active participants within the perceptual process, and connect with our sounds to build their own meanings. Embracing this complexity, and audience subjectivity, liberates us from narrow fixation on tools or equipment, and instead focuses our attention onto the external human listener and our shaping of their listening experience. It does not matter what tool or technique you use, it only matters that the audience connects with the sound that you craft.

6.5.4 *Attend to Creation for 'the other' Listener*

We develop Electroacoustic Compositions and Film Soundtracks to communicate to audiences, and thus it is essential that an openness to 'the other' is featured as a primary driver within our understanding of sonic practices. This demands an active consideration of 'other' listeners – respecting people's potential unique interpretations and opinions – and embracing this potential of plurality in order to powerfully inform creative decisions. Such an approach includes being able to appreciate diverse perspectives and not simply working to a homogenised minority mindset. Such an approach is encapsulated in our technique of Listening Empathy *(Techniques),* an active approach to embrace and understand the other via demonstrating receptivity to sounds potential to communicate emotion and affective impressions.

6.5.5 (Film) Embrace the Potential of Sonic Creativity at the Start of the Filmmaking Process

In film, sonic sensibilities have the potential to modulate and inform all aspects of filmmaking. Sound has the potential to actively inform and shape decision-making in deep network with all of the other creative and practices that make up the collaboration of film. Sound is not a final layer to be 'iced on', but a potential equal participant in the creative conversation and development process. What new approaches to storytelling might be accessed if we let sound lead?

By demonstrating sonic creativity to be an unfolding iterative Process of Discovery (Principles), we build a stronger argument to include and engage sound much earlier within the collaborative filmmaking process. Sound has the potential to respond dynamically to the creative facets of practice in complex ways, and to modulate the act of storytelling in a foundational way, opening new opportunities and new insights as part of the collaborative creative conversation.

Our Map of Sonic Creativity might provide various starting points for exploration and enquiry for filmmakers to help demystify sounds' potential. It foregrounds creative decision-making over the technical, challenges preconceptions around what sound is and does, and might enable creative conversations between Filmmakers working across image and sound, fostering greater understanding of the contribution that sound can make.

6.5.6 Open Access to Voices who have been Less Represented in the Past

As explored in Chapter 1 (Section 1.4.3, p.12–14.), the statistics of engagement with Film Sound, academic Music Technology, and Music Production practices paint a picture of a largely heterogeneous population of limited diversity. Such features are symptoms of complex problems of access and engagement, but it is clear that the status quo risks reinforcing a limited way of thinking about how to do sound practice. This is not to deny the white male standpoint entirely, it is to simply recognise that this should not be regarded as the universal view.

New practitioners bring new perspectives, and we must be open to (and continue to open even further to) a multiplicity of voices. This is not only about benefit to others. In recognising the alternative approaches of others we become more open to learn about our own practices. Thus embarking upon continually evolving processes of Constant Enquiry, which leads both to a stronger field and more diverse aesthetic opportunities for all.

We believe that a shift in approach allowing for non-prescriptive conceptions of practice (across both Electroacoustic Music and Film Sound), can empower a more equitable access to sound practices and enable the entrance of more diverse voices to contribute to the field. Such a move would have overall positive benefits for the aesthetic development and increasing richness of practices across both fields.

Wherever absolutes exist, claims can surface for possession of the 'true and correct answer'. Such claims are instantly exclusionary, situating a division between those apparently in the know and those outside. Such a position is a fallacy. If we fail to tackle such implicit bias and informal forms of exclusion, we are limiting the potential of our field to benefit from broader insights and valuable contributions from more diverse practitioners. Celebrating that there are manifold ways of imagining and doing sound practice, means that we can welcome the full richness of situated knowledge, embracing new inspirations and ideas, all the while sidestepping forms of gatekeeping which limit all of us.

6.6 The Journey So Far…

In this book we set out to better understand creative processes of making across both Film Sound and Electroacoustic Music. Through listening to world leading practitioners talk about their processes we have revealed underlying *Principles*, *Approaches,* and *Techniques* which drive their working practices, recognising and celebrating the complexity and inherent creativity within these arts of sound.

By engaging in extensive discussions around the nature of creative practice in sound, we have developed a rich resource that exceeds our hypotheses. Celebrating creative process as a rich, complex, and valid form of knowledge, our research has attempted to provide waypoints to inspire, drive, and empower opportunities for ever greater sonic creativity. Whether you have been reading as a sound practitioner seeking to reflect on your own creative practices, a potential collaborator looking to know more about what sound can do (Filmmaker, Cinematographer, Scriptwriter etc.), or a student studying towards a career in sound or music; we hope that our investigation has provided you with an opportunity to unlock a deeper insight into the complex potential of sound.

We hope that this book will contribute to the rich world of sonic creativity by encouraging appreciation for the diverse possibilities available. In appreciating multiplicity and potential as core features of the creative process, we are better informed to enable more open and honest exchanges about practice; to centre our attention on the human aspects of practice (rather than the technical); to open access to historically less represented voices; to provide collaborative exchange across and between the artfroms of Film Sound and Electroacoustic Music, and provide new routes of access for those who could greatly benefit from a fuller understanding of the potential of sound as an artistic medium and collaborative partner. This celebration of working *with* sound highlights the inherrent decision making processes which are fundamentally creative, and thus situates our practices as far more than simple technical actions but as genuine artistic contributions. We hope that, as a result of our study, Filmmakers, Producers, Directors, Designers might develop greater recognition for the possibilities that sound provides when engaged as an equal partner in the creative process.

Insights into creative practice and creativity will also be valuable for practice-research scholars, those seeking to learn more about creative practices, and those engaged in PhD study around creative practice, whether in sound or other artforms. We hope those engaged in such will embrace with confidence the richness of subjectivity with the support of our analysis, and utilise Our Map of Sonic Creativity as a model for wider and future understandings of making and doing creative practice. Though developed out of a detailed study of sonic practices, the findings have the potential to be transferable to other creative practices and we look forward to learning about their possible future application in a range of contexts.

We originally set out to identify any similarities or differences that existed between the creative approaches applied within Film Sound and Electroacoustic Music, imagining that we might end up with results that encouraged a comparative analysis (built around differences). However, the deeper we reflected and the more layers we peeled back, the more we discovered that responses pointed towards a set of common drivers of creativity. As a result, rather than the impetus for insight being difference, it has actually been shared commonality that has been the key to unlocking the most valuable insights from this project.

Film Sound and Electroacoustic Music practice are independent artforms with their own distinct traditions, tropes, aesthetic remits, and goals. Each field maintains a sense of its own identity and this identity defines a certain scope of expectation for what constitutes a work within each field. But, such definitions and identities are not absolute and fixed. They have the potential to modulate, develop, and change. That we have revealed an underlying similarity in aesthetic sensibility between these practices – leading to Our Map of Sonic Creativity and the framework of shared *Principles*, *Approaches* and *Techniques* it is built upon – suggests that there is significant ongoing and future potential for collaboration and sharing between these practices, and expansion of this should be an imperative for ongoing research in the future.

6.7 Closing Thoughts

In bringing together practitioners from these different but contiguous worlds, we have been able to reflect upon what it means to create with sound and to identify assumptions and boundaries which have formed around each field. With a shared passion for sound, artistic expression, and so many complementary approaches, our collision of Film Sound and Electroacoustic Music did not act to draw these practices/practitioners closer together, but simply demonstrate how thin was the veil that previously separated them.

Our focus on process means that we inherently welcome future engagement, and see this work as a single contribution to a continually unfolding dialogue between practitioners and researchers. In sharing our research process, the methods we applied and the steps undertaken, we hope to encourage further conversations and follow-on research projects which continue to build, celebrate, and critically reflect upon practices with sound. In creating these conversations we have stimulated dialogues, exchanges and networks that have begun to venture pathways of partnership across and between these diverse fields. The project has demonstrated that there is much that we can learn from one another when we engage in dialogue. None of this would have been possible without the generous contributions of our participants, who all openly shared their approaches, enabling us to de-mystify and make less abstract some of the uncertainties and intangibilities of creative practice.

Sound is always part of something bigger. A larger context which engages not just materials and technology, but people. Those people who engage, articulate, and form communities around their shared interests are the most powerful asset within any sonic practice. It has been a pleasure of this project to engage with others, joining with them to talk about sound, to be humbled by their talent and generosity in sharing their practices, and to be able to bring these contributions together in this book, making such noises that highlight the rich artistry at play within both Film Sound and Electroacoustic Music.

Notes

1 Each of the points emerged from a process of triangulation and detailed cross-referencing between professional responses to questions of creative practice. These factors might not have been explicitly declared, but were implicit in statements made by practitioners.

2 An example of this is that, any different creative given the same materials might have worked on them in an entirely different way.

3 The creative act is not a static process. We cannot stop or segment time, so we must embrace this as movement constantly in flux.

4 As the phenomenologist Maurice Merleau-Ponty asserts, the very act of 'perception becomes an 'interpretation' of the signs that our senses provide in accordance with the bodily stimuli, a 'hypothesis' that the mind evolves to explain its impressions to itself' (Merleau-Ponty 1962, p.39).

5 The simplistic notion of the great creative vision is what Haraway might term "a full and total position", and she argues that the search for such "is the search for the fetishized perfect subject of oppositional history" (1991, p.198).

6 Such 'shared conversations' (Haraway 1991) are possible within the complex networks of connection binding people together, producing meeting points of shared solidarity around which layers of difference nevertheless persist.

7 One of the questions on our key 'schedule of topics' (discussed in Chapter 2) directly solicited responses to this question: "Is your creative process pre-planned or emergent?". Many of our conversations reflected upon this topic, providing a rich discourse of responses from which we were able to glean insight.

8 One sound may be perceived as all or none of – diegetic, musical, real – heard differently by each listener.

9 As Haraway identifies, "only those occupying positions of the dominators are self-identical, unmarked, disembodied, unmediated, transcendent" (Haraway 1991, p.193), which is to say that those who seek to enforce limited rules and universalisation are those in established situations of power seeking to subjugate and retain control over others.

10 For example: insight into where the original source sound came from, how it was developed and what tools and techniques were employed in rendering its final result.

11 This oversimplification of process is used extensively by hardware and software manufacturers in their marketing campaigns, who proffer quick solutions to complex creative challenges via their tools.

List of Referenced Projects and Art Works

Music

- *Aeolian* (2019) – Annie Mahtani – pages 151, 152
- *Attending to Sacred Matters* (2002) – Hildegard Westerkamp – page 168
- *A Walk Through the City* (1992) – Hildegard Westerkamp – pages 126, 168
- *Beneath the Forest Floor* (1992) – Hildegard Westerkamp – page 170
- *Claustro* (2019) – Nikos Stavropoulos – page 157
- *Continual* (2020) – KMRU – page 181
- *Cricket Voice* (1987) – Hildegard Westerkamp – pages 128, 171
- *Encounters in the Republic of Heaven* (2014) – Trevor Wishart – pages 48, 57, 141, 173
- *Fabulous Paris: A Virtual Oratorio* (2007) – Trevor Wishart – page 173
- *From the India Sound Journal* (1993–2000) – Hildegard Westerkamp – page 128
- *Globalalia* (2014) – Trevor Wishart – pages 57, 141, 142
- *Inversions* (2015) – Annie Mahtani – page 190
- *Karst Grotto* (2017) – Nikos Stavropoulos – page 157
- *Kits Beach Soundwalk* (1989) – Hildergard Westerkamp – pages 127–128
- *Lamentations* (2010) – John Young – page 207
- *Moments of Laughter* (1999) – Hildegard Westerkamp – pages 125–126
- *NUBI* – KMRU – page 149
- *Once He Was a Gunner* (2020) – John Young – page 137
- *Past Links* (2008) – Annie Mahtani – page 152
- *Racines Tordues* (2019) – Annie Mahtani – pages 190, 192–193
- *Ricordiamo Forlì* (2005) – John Young – pages 38, 134, 137, 138
- *Round Midnight* (2019) – Annie Mahtani – page 193
- *Red Bird* (1978) – Trevor Wishart – page 178
- *Tao* (1984) – Annette Vande Gorne – page 130
- *Topophilia* (2016) – Nikos Stavropoulos – page 157
- *The Garden of Earthly Delights* (2020) – Trevor Wishart – page 173
- *Tongues of Fire* (1994) – Trevor Wishart – pages 57, 141
- *Two Women* (2000) – Trevor Wishart – page 173
- *Whisper Study* (1975) – Hildegard Westerkamp – pages 40, 124–125

Film

- *Bohemian Rhapsody* (2018) – pages 82, 84–85, 89, 179
- *Bohemian Rhapsody* (1975) – pages 82, 84–85, 89, 179

DOI: 10.4324/9781003163077-7

TV Series

Sound Recordings

Computer Games

References

And Breathe. (2021). Directed by Sima Gonsai [Dance Film]. UK: Sima Gonsai Films.

Avarese, J. (2017). *Post Sound Design: The Art and Craft of Audio Post Production for the Moving Image.* New York: Bloomsbury.

Baby Daddy. (2018). Directed by Elinor Coleman [Theatre]. UK: Birmingham Repertory Theatre.

Barton Fink. (1991). Directed by Joel Coen [Film]. US: 20th Century Fox.

Bohemian Rhapsody. (2018). Directed by Bryan Singer [Film]. US: 20th Century Fox.

Born, G. (1995). *Rationalizing Culture: IRCAM, Boulez, and the Institutionalization of the Musical Avant-Garde.* Berkeley: University of California Press.

Born, G. (2005). *Uncertain Vision: Birt, Dyke and the Reinvention of the BBC.* New York: Vintage.

Born, G. (2010). Social and the Aesthetic: For a Post-Bourdieuian Theory of Cultural Production. *Cultural Sociology*, 4(2), pp.171–208.

Born, G., & Devine, K. (2015). Music Technology, Gender, and Class: Digitization, Educational and Social Change in Britain. *Twentieth-Century Music*, 12(2), pp.135–172.

Born. G., & Hesmondhalgh, D. (eds.). (2000). *Western Music and Its Others: Difference, Representation, and Appropriation in Music.* Berkeley: University of California Press.

Bresson, R. (1986). *Notes on the Cinematograph.* Trans. Jonathan Griffin. New York: New York Review of Books.

Burt, T. W. (1996). Some Parentheses Around Algorithmic Composition. *Organised Sound*. Cambridge University Press, 1(3), pp.167–172. https://doi.org/10.1017/S1355771896000234.

Butt, E. (2020). Diversity in Post Production Sound Roles in UK Television Production. Sir Lenny Henry Centre for Media Diversity. Birmingham City University. Available from: https://bcuassets.blob.core.windows.net/docs/diversity-in-post-production-sound-roles-in-uk-television-production-132514656878973963.pdf (Last Accessed: 05 December 2022).

Chicago PD. (2014 – present) [TV Series]. US: NBC Universal Syndication Studios.

Chion, M. (1994). *Audio-Vision: Sound on Screen.* 2nd edition. New York: Columbia University Press.

Cinema Audio Society. (2020). Sound Credit Initiative put forth by CAS, MPSE, & AMPS. Available from: https://cinemaaudiosociety.org/cinema-audio-society-joins-together-with-motion-picture-sound-editors/ [Last Accessed: 05 March 2023].

Clifford, J., & Marcus, G. E. (1986). *Writing Culture: The Poetics and Politics of Ethnography.* In James Clifford and George E. Marcus (eds.). Berkeley: University of California Press.

Cox, C., & Warner, D. (2002). *Audio Culture: Readings in Modern Music.* New York: Continuum.

Deleuze, G., & Parnet, C. (1987). *Dialogues.* Trans H. Tomlinson and B. Habberjam, London: The Athlone Press.

Descartes, R. (1637). *Discourse on Method.* 1911 Trans. Edited by S. Haldane and G. R. T. Ross; first five parts. Cambridge: Cambridge University Press.

Donnelley, K. (2005). *The Spectre of Sound: Music in Film and Television.* London: Bloomsbury.

Dune. (1984). Directed by David Lynch [Feature film]. US: Universal Pictures.

Elephant. (2003). Directed by Gus Van Sant [Feature film]. Burbank, CA: Fine Line Features, & New York: HBO Films.

Emmerson, S. (2001a). From Dance! To "Dance": Distance and Digits. *Computer Music Journal,* 25(1), pp.13–20.

Emmerson, S. (2001b). *Pentes: A Conversation with Denis Smalley. SAN Diffusion.* Available at: https://electrocd.com/en/presse/428_5 (Accessed: 1 September 2022).

Everest. (2015). Directed by Baltasar Kormákur [Film]. US: Universal Pictures.

Fahim, H. (ed.). (1982). *Indigenous Anthropology in Non-Western Countries.* Durham, NC: Carolina Academic Press.

Fantasia. (1940). Directed by Armstrong, S., Algar, J., Roberts., B., Satterfield, P., Sharpsteen, B., Hand., D., Luske, H., Handley, J., Beebe, F., Ferguson, N., Jackson, W. California, US: Walt Disney Productions.

Feld, S. (1987). Dialogic Editing: Interpreting how Kaluli Read Sound and Sentiment. *Cultural Anthropology,* 2(2): pp.190–210.

Follows, S. (2019). *How Often Are Women Hired in Key Film Departments?* Available at: https://stephenfollows.com/women-in-key-film-departments/ (Accessed: 5 April 2021).

Furley, S. (2011). *Ann Kroeber Special: A Pioneering Sound Woman.* Available at: https://designingsound.org/2011/10/07/ann-kroeber-special-a-pioneering-sound-woman/ (Accessed: 10 July 2021).

Game of Thrones. (2011–2019) [TV Series]. US, Warner Bros. Television Distribution.

Gravity. (2013). Directed by Alfonso Cuarón [Film]. California, US: Warner Bros. Pictures.

Hanson, H. (2017). *Hollywood Soundscapes.* London, UK: Bloomsbury Publishing.

Haraway, D. J. (1989). *Primate Visions: Gender, Race, and Nature in the World of Modern Science.* New York: Routledge.

Haraway, D. J. (1991). *Simians, Cyborgs, and Women: The Reinvention of Nature.* New York: Routledge.

Harrison, J. (1999). *Diffusion: Theories and Practices, with Particular Reference to the BEAST System.* Montréal: Communauté électroacoustique canadienne/Canadian Electroacoustic Community. eContact!, 2.4 Multi-Channel Diffusion.

Hill, A. (2013). Interpreting Electroacoustic Audiovisual Music. (PhD thesis) De Montfort University.

Hill, A. (2017). Listening for Context – Interpretation, Abstraction & The Real. *Organised Sound,* 22(1), pp.11–19.

Holloway. L. (2011) Donna Harraway, In P. Hubbard & R. Kitchin (ed.) *Key Thinkers on Space and Place.* 2nd edition. pp.219-226. London: Sage.

Ingold, T. (2000). *The Perception of the Environment.* London: Routledge.

Ingold, T. (2007). 'Against soundscape'. In A. Carlyle (ed.) *Autumn Leaves: Sound and the Environment in Artistic Practice,* pp.10–13. Paris: Double Entendre & CRISAP.

Ingold, T. (2013). *Making: Anthropology, Archeology, Art and Architecture.* London: Routledge.

KMRU. (2020). Continual. *Continual.* [EP]. Available at: https://kmru.bandcamp.com/album/continual (Accessed: 1 August 2021).

Knight-Hill, A. (2019). Sonic Diegesis: Reality and the Expressive Potential of Sound in Narrative Film, *Quarterly Review of Film and Video,* 36(8), pp.643–665.

Knight-Hill, A. (2020a). 'Audiovisual Spaces: Spatiality, Experience and Potentiality in Audiovisual Composition.' In Knight-Hill (ed.) *Sound & Image: Aesthetics and Practices.* London: Routledge.

Knight-Hill, A. (2020b). 'Electroacoustic Music: An Art of Sound.' In Filimowicz (ed.) *Foundations of Sound Design for Linear Media.* London: Routledge.

Ladefoged, P., & Maddieson, I. (1995). *The Sounds of the World's Languages – Phonological Theory.* Hoboken, NJ: Wiley.

Langer, S. (1957). *Philosophy in a New Key: A Study in the Symbolism, of Reason Rite and Art.* Cambridge, MA: Harvard University Press.

Law and Order. (1990–2010, 2022 – present). [TV Series] US: National Broadcasting Company.

Lazarus, A. (1974). The Use of the FRAP (Flat Response Audio Pickup) in Professional Recording. *AES Convention* 1974. Available from: https://www.aes.org/e-lib/browse.cfm?elib=2539 (Last Accessed: 04 December 22).

LeCompte, M., & Goetz, J. (1982). Problems of Reliability and Validity in Ethnographic Research. *Review of Educational Research.* Spring 1982, 52 (1), pp. 31–60.

Lovecraft Country. (2020). New York, US: HBO.

Mahtani, A. (2015). *Inversions.* Available at: https://soundcloud.com/annie-mahtani/inversions-2015-stereo-mixdown (Accessed: 3 November 2022).

Mahtani, A. (2019a). Aeolian. *Racines* [CD]. Canada: Empreintes DIGITALes.

Mahtani, A. (2019b). Past Links. *Racines* [CD]. Canada: Empreintes DIGITALes.

Mahtani, A. (2019c). Racines Tordues. *Racines* [CD]. Canada: Empreintes DIGITALes.

Mahtani, A. (2019d). 'Round Midnight. *Racines* [CD]. Canada: Empreintes DIGITALes.

Margetson, E. (2021). *Sonic Immersion: Reaching New Audiences through Sound.* (PhD thesis) University of Birmingham.

Marstal, H. (2020). 'Slamming the Door to the Recording Studio - Or Leaving it Ajar?' In R. Hepworth-Sawyer, J. Hodgson, L.King, M. Marrington (eds.) *Gender in Music Production.* New York: Routledge.

Merleau-Ponty, M. (1962). *Phenomenology of Perception.* London: Routledge.

MK Ultra. (2018). Directed by Rosie Kay Dance Company [Dance]. UK: Rosie Kay Dance Company.

Mowgli: Legend of the Jungle. (2018). Directed by Andy Serkis [Film]. California, US: Netflix, Inc.

Mufasa: The Lion King. (2024). Directed by Barry Jenkins [Film]. US: Disney +.

Nielsen, T. (2011). *On the Art of Economy.* Available at: https://designingsound.org/2011/08/16/tim-nielsen-special-on-the-art-of-economy/ (Accessed: 7 November 2022).

Normal People. (2020). Directed by Lenny Abrahamson & Hettie Macdonald [TV series]. UK: BBC Studios.

O'Callaghan, J. (2011). Soundscape Elements in the Music of Denis Smalley: Negotiating the Abstract and the Mimetic. *Organised Sound,* 16(1), pp.54–62.

Ohnuki-Tierney, E. (1984). "native" anthropologists. *American Ethnologist,* 11(3), pp.584–586.

One Chicago. (2012–2022) [TV Series]. US: National Broadcasting Company.

Once Upon a Time in America. (1984). Directed by Leone, S. United States: Warner Bros and Italy: Titanus.

Once Upon a Time in the West. (1968). Directed by Sergio Leone. United States: Paramount Pictures and Italy: Euro International Films.

Plato. (1943). *Plato's The Republic.* Translated by Benjamin Jowett. New York: Books, Inc.

Queen. (1975). Bohemian Rhapsody. *A Night at the Opera* [CD]. United Kingdom: EMI.

Queen Official. (2018). *Put Me In Bohemian!.* Upload 29 June. Available at: https://www.youtube.com/watch?v=aCDUSjR0uf0 (Accessed: 3 July 2022).

Romeo and Juliet. (2021). Directed by Rosie Kay Dance Company [Dance]. UK: Rosie Kay Dance Company.

Room. (2015). Directed by Lenny Abrahamson [Film]. Canada: Elevation Pictures, UK and Ireland: StudioCanal, United States: A24.

Rogers, H. (2014). *Music and Sound in Documentary Film.* London: Routledge.

Schafer, R. M. (1994). *The Soundscape: Our Sonic Environment and the Tuning of the World.* Rochester: Destiny Books.

Schopenhauer, A. (1819). *The World as Will and Representation.* New York: Dover Publications Inc.

Smalley, D. (1986). 'Spectro-morphology and Structuring Processes.' In S. Emmerson (ed.) *The Language Electroacoustic Music.* London: Palgrave Macmillan.

Smalley, D. (2007). Space Form and the Acousmatic Image. *Organised Sound,* 12(1), pp.35–58. London: Palgrave Macmillan.

Smalley, D. (2010). 'Spectromorphology, Motion and Meta-motion in Denis Smalley's 'Vortex'.' In E. Gayou and F. Couture (eds.) *Polychrome Portraits,* pp.89–101. Paris: INA-GRM.

Sonnenschein, D. (2004). *Sound Design: The Expressive Power of Music and Sound Effects in Film*. Studio City, CA: Michael Wiese Productions.

Sporton, G. (2015) *Digital Creativity: Something from Nothing*. London: Palgrave Macmillan.

Stacy, L., Smith, K. P., Clark, H., Case, A., & Choueiti, M. (2020). *Inclusion in the Recording Studio? Gender and Race/Ethnicity of Artists, Songwriters & Producers across 800 Popular Songs from 2012–2019*. Available at: https://assets.uscannenberg.org/docs/aii-inclusion-recording-studio-20200117.pdf (Accessed: 4 November 2022).

Star Wars: Episode IV A New Hope (1977). Directed by George Lucas. [Film] US: 20th Century Fox.

Stavropoulos, N. (2016). *Topophilia*. Available at: https://soundcloud.com/nikos-stavropoulos/topophilia (Accessed: 3 November 2022).

Stavropoulos, N. (2017). *Karst Grotto*. Available at: https://soundcloud.com/nikos-stavropoulos/karst-grottoexcerpt-stereo-reduction (Accessed: 3 November 2022).

Stavropoulos, N. (2018). *Inside the Intimate Zone: The Case of Aural Micro-Space in Multichannel Compositional Practice*. In: Proceedings of the 15th Sound and Music Computing Conference (SMC2018). Sound and Music Computing Network, pp.113–117. ISBN 978–9963-697-30–4.

Stavropoulos, N. (2019). *Claustro*. Available at: https://soundcloud.com/nikos-stavropoulos/claustro-stereo-reduction (Accessed: 3 November 2022).

Sterne, J. (2003). *The Audible Past*. Durham, NC: Duke University Press.

Strathern, M. (2004). *Partial Connections*. Oxford: Altamira Press.

Tattoo Stories. (2021). Directed by Steve Johnstone. [Podcast]. Smethwick: Black Country Touring.

The Conversation. (1974). Directed by Hackman, G., Coppola, F. F., Cazale, J., Forrest, F., Williams, C., Roos, F., & Shire, D. Hollywood [Film]. California: Paramount Pictures Corp.

The Killing of Two Lovers. (2021). Directed by Robert Machoian [Film]. US: Neon.

The Little Stranger. (2018). Directed by Lenny Abrahamson [Film]. France: Pathé Distribution and UK and Ireland: 20th Century Fox.

The Lord of the Rings: The Rings of Power. (2022). Directed by Payne, J. D., and McKay, P. California [TV Series]. US: Amazon Studios.

The Shining. (1980). Directed by Stanley Kubrick. [Film]. US, Warner Bros. & UK, Columbia-EMI-Warner Distributors.

The Underground Railroad. (2021). Directed by Barry Jenkins [Drama Series]. US, Amazon Prime Video.

Turchi, P. (2004). *Maps of the Imagination: The Writer as Cartographer*. San Antonio, TX: Trinity University Press.

Vande Gorne, A. (1984). Eau. *Tao* [CD]. Canada: Empreintes DIGITALes.

Vande Gorne, A. (2017). *Treatise on writing acousmatic music on fixed media*. Ohain, LIEN vol. VIII, 2017 80p. (first French edition), and vol. IX, 2018 (English translation), pp.14–23.

Varèse, E. (2002). "The Liberation of Sound." In C. Cox and D. Warner (eds.) *Audio Culture: Readings in Modern Music*, pp.17–21. London: Bloomsbury.

Vlad, G. (2017a). *2017 Sound recording expedition to Senegal*. Available at: https://youtu.be/C4sLlLWjDf0 (Accessed: 1 August, 2021).

Vlad, G. (2017b). Field recording trip to South Africa – 2016. Available at: https://mindful-audio.com/blog/2017/9/4/field-recording-trip-to-south-africa-2016 (Accessed: 1 August, 2021).

Vlad, G. (2018). *Thunderstorm At Langoue Bai*. Available at: https://soundcloud.com/georgevlad/thunderstorm-at-langoue-bai?in=georgevlad/sets/congo-basin-rainforest-2018 (Accessed: 1 August, 2021).

Vlad, G. (2020). *Intense Dusk Chorus In The Borneo Rainforest*. Available at: https://soundcloud.com/georgevlad/dusk-borneo (Accessed: 1 August, 2021).

Vlad, G. (2021a). *Chromatic Rain in the Namib Desert – Nature Sounds*. Available at: https://youtu.be/P09ZwPBwenM (Accessed: 1 August, 2021).

Vlad, G. (2021b). *The Weird Sound of Empress cicadas*. Available at: https://youtu.be/6ZPDYlvj7MM (Accessed: 1 August, 2022).

'Walk of Punishment'. (2013). *Game of Thrones,* Season 3, Episode 3. Directed by David Benioff. HBO, April 14, 2013.

Warren, S. (2021). *AFEM Gender Diversity in the Electronic Music Industry Survey.* Available at: https://associationforelectronicmusic.org/2021/12/20/afem-diversity-inclusion-working-group-present-the-gender-diversity-in-the-electronic-music-industry-report/ (Accessed: 1 November 2022).

Westerkamp, H. (1975). Whisper Study. *SFU 40* [CD]. Burnaby, Canada: Simon Fraser University.

Westerkamp, H. (1987). Cricket Voice. *Transformations* [CD]. Canada: Empreintes DIGITALes.

Westerkamp, H. (1989). Kits Beach Soundwalk. *Transformations* [CD]. Canada: Empreintes DIGITALes.

Westerkamp, H. (1992a). A Walk Through the City. *Transformations* [CD]. Canada: Empreintes DIGITALes.

Westerkamp, H. (1992b). Beneath the Forest Floor. *Transformations* [CD]. Canada: Empreintes DIGITALes.

Westerkamp, H. (1993–2000). *From the India Sound Journal.* Available at: https://www.hildegardwesterkamp.ca/sound/comp/2/indiasj/ (Accessed: 1 September 2021).

Westerkamp, H. (1999). Moments of Laughter. *Breaking News* [CD]. Canada: earsay music

Westerkamp, H. (2002). Attending to Sacred Matters. *Into India* [CD]. Canada: earsay music.

Wilkinson, A. (2014). *A Voice from the Past.* Available at: https://www.newyorker.com/magazine/2014/05/19/a-voice-from-the-past (Accessed: 30 November 2021).

Wishart, T. (1978). Red Bird. *Red Bird (A Political Prisoner's Dream)* [Vinyl]. York: York Electronic Studios.

Wishart, T. (1994a). *Audible Design: A Plain and Easy Introduction to Practical Sound Composition.* York: Orpheus the Pantomime Ltd.

Wishart, T. (1994b). Tongues of Fire. *Tongues of Fire* [CD]. UK: Orpheus the Pantomime Ltd.

Wishart, T. (1996). *On Sonic Art.* 2nd edition (Edited by Simon Emmerson). Routledge: London.

Wishart, T. (2000). Two Women. *Voiceprints* [CD]. New York: EMF Media.

Wishart, T. (2007). *Fabulous Paris: A Virtual Oratorio.* Available at: https://www.discogs.com/release/1492681-Trevor-Wishart-Fabulous-Paris-A-Virtual-Oratorio.

Wishart, T. (2014a). Globalalia. *Globalalia/Imago* [CD]. UK: Orpheus The Pantomime.

Wishart, T. (2014b). Encounters in the Republic of Heaven. *30 Jahre Inventionen VII 1982–2012* [CD]. Germany: Edition RZ.

Wishart, T. (2020). The Garden of Earthly Delights. *The Garden of Earthly Delights* [CD]. London: ICR Distribution.

Witts, R. (2004). 'Stockhausen vs. the 'Technocrats'.' In C. Cox and D. Warner (eds.) *Audio Culture: Readings in Modern Music,* pp.381–385. New York: Continuum.

Wolfenstein II: The New Colossus (2017). Microsoft Windows, PlayStation 4, Xbox One and Nintendo Switch. Bethesda, MD: Bethesda Softworks.

Women's Audio Mission. (2022). *About Page.* Available at: https://womensaudiomission.org/about/ (Accessed: 10 October 2022).

Young, J. (2007). Ricordiamo Forlì (2005). On *Lieu-temps.* [DVD-A]. Montréal: Empreintes DIGITALes.

Young, J. (2010). Lamentations, on *Metamorphose 2010* [CD], Ohain: Musiques & Recherches.

Young, J. (2020). Once He Was a Gunner (2020), on *Histoires des soldats,* Montréal: Empreintes DIGITALes.

10 Soldiers. (2019). *Directed by* Rosie Kay Dance Company [Dance]. UK: Rosie Kay Dance Company.

Index

Note: *Italic* page numbers refer to figures and page numbers followed by "n" denote endnotes.